Sugar Trains Pictorial

Comments on the author's previous volume, *Sugar Trains: Narrow Gauge Rails of Hawaii:*

"A tremendous job of putting material together, the result is a valuable addition to Hawaiiana."

Archives of Hawaii

"Reads more like a historical novel than a book on railroads. Heavily researched; truly a valuable book."

Alexander & Baldwin Inc.
Ampersand &&&

"A superb book on a subject which we would have believed to be non-researchable."

DeGolyer Foundation Library
Everett L. DeGoyler, Jr.

Glenwood Publishers
Post Office Box 194
Felton, California 95018

Photo:
Iolani Palace, Honolulu, Hawaii, circa 1883. Built by Kalakaua, Rex. In the vernacular — Bird of Heaven *literally* — Heavenly Hawk *as the high flight of the Io bird symbolized royalty.*

Sugar Trains Pictorial

by Jesse C. Condé

Printed and bound in the United States of America
by Big Trees Press and Cardoza-James
First Edition
ISBN 0-911760-21-0
Library of Congress Catalog Card No. 75-11103

Dedicated
To
The Memory
Of
Nelson Prather
Dean of Public Relations Men of Hawaii
1912 - 1971

(Infinitely intrigued with the idea that a Flight Engineer, assigned to heavy jet aircraft, would be so interested in the Hawaiian plantation rail scene, Nelson Prather, "Mr. P.R. Man of Hawaii", was sufficiently attracted as to allow a thumb nail sketch of the author's first Hawaiian work—"Narrow Gauge in a Kingdom—The Hawaiian Railroad Company", to be reported. The vehicle was the July/August 1971 issue of "Ampersand &&&", the house organ of Alexander & Baldwin, Inc., and it included data on the old Kahului Railroad Company, once controlled by A&B.)

KINAU
H.R.R. Co.
Kotowski

Contents

Drawing an explosion protection car and trailing a string of the so-called "Kalakaua Cars," Kinau, *#1 locomotive of the old Hawaiian Railroad Company, crosses one of the low trestles in North Kohala in 1882. (Michael Kotowski from an Archives of Hawaii photo, 1880s.)*

Contents

Preface

These photographs, which continue the heritage of the Hawaiian Plantation railroads and their artifacts, could not be crowded into the allotted pages of the original volume even though the Editor strove to his utmost to accommodate them. Of all remaining from the original *Sugar Trains: Narrow Gauge Rails of Hawaii*, many found their way, so to speak, to the cutting room floor.

Since several previous "photographic histories" of plantation railroads, covering a span from 1879 until 1959, had been lost or inadvertently destroyed, it was felt that these remainders should not also be lost. The result is this *Sugar Trains Pictorial.*

At the start of the exercise to uncover the record, few photographs of positive identity could be found. As time passed, however, they were gradually located in many and varied places. Indeed, after the publication of the original work many more photos were proferred by interested rail buffs and from other sources close to the Hawaiian scene.

Herein are also included some photographic history of seven Hawaiian Island common carriers. Certainly a class by themselves, they are included here since their origin was actually based on the acute need to provide a vehicle for moving the great sugar production of the Hawaiian enterprise of the times. So inter-mixed with the plantation carriers were they that it is often hard to separate their actual aims. Some carried passengers and pineapple, but for their creation to transport sugar neither of the other revenue sources could have benefited as they did.

Without this means of transport, available from the monetary returns provided with the advent of the Treaty of Reciprocity, it is possible that what was to be the greatest industry of the Islands could not have prospered to the point where it became necessary to resort to the internal combustion engine, supplanting the railroads with Tournahaulers and giant trucks to transport the masses of sugar cane ultimately produced in Hawaii.

All have been combined here. And with these additions to the railroad scene in Hawaii it is hoped that their great contribution to Hawaiiana will not be forgotten.

My sincere "Mahalo" to those who have contributed so very generously.

Aloha, nui loa,
Jesse C. Condé

Foreword

Transport has always been a major impediment to Hawaiian sugar planters. Railways were one solution in the rough burnt fields, as locomotives hauled endless lines of cars over uneven tracks that went forward with the harvesters.

Sugarcane is the bulkiest of the world's crops. At maturity there may be 30,000 stalks up to 30 feet long on an acre. In the heyday of plantation railroads yields of from 40 to 70 tons of cane per acre had to be harvested and transported as fast as possible to the factories. Speed was of the essence both to avoid sugar losses after cutting and to keep the voracious mills fed. Hawaiian sugar factories operated twenty-four hours a day for at least ten months of every year. Supplies of harvested cane had to reach the manufacturing machinery in exactly the right amounts every hour of every day. A sufficient supply of cane had to be garnered during the working hours of the day to keep the factory machinery running at full efficiency, not only while harvesting was actually going on but for the rest of the afternoon and night.

From the earliest days of the sugar industry in Hawaii the major consideration has been the need of doing much with limited labor. The supply of workers imported with difficulty and great expense to the Islands was barely enough and at no time were there superfluous numbers.

The "gently rising land" observed by Captain Cook is really comprised of deep gulches, steep grades and on the whole an exceedingly uneven terrain. Most plantation areas generally follow the line of the coast and extend inland on the slopes. Often the grade is extremely steep. Cane was hand cut by sweaty, grimy men laboring their utmost to harvest the almost impenetrable fields of sugarcane. The distances that cane had to be carried from the centers of the fields to the factories and the necessity for maintaining a steady procession of cane were overcome by ingenuity and innovation.

Railways were introduced on sugar plantations in the 1870s and the lines were rapidly extended. The planters generally laid their railway tracks crosswise of the plantation, parallel to the shoreline, and if possible midway between the high and low fields. The distance that cane had to be fetched up to the line or down to it was thus not too great.

The ponderous and labor-wasting job of getting cane out of the centers of the fields was reduced by laying portable tracks. Along the permanent way frequent switches were provided and tracks were laid from these, being extended as the cane was cut. Empty cars were hauled on the portable tracks close to the site of each hour's harvest; filled cars were backed onto the main line and moved to the factory. At the factory small locomotives constantly shunted the filled cars into position to be unloaded and empties were moved to the edge of the yard to be dispatched again.

Infinite forethought went into all of these operations. Fields seeded twenty-odd months before were scheduled to the appropriate moment for harvest; adjoining areas had to be ready for successive harvest. Costly transportation equipment had to be kept steadily at work so the factories were never waiting.

The railway, like so many of the last century's innovations, was a perfect complement to the times. Hawaii needed stable transport to continue to operate and the railroad supplied this need which helped bring about the industry's growth. Much of the hinterland of the various islands was opened up and put into production of sugar. With manpower, mules and oxen this could not have been done profitably, but the simplicity, durability and economy of the railroads helped accomplish the great development of the sugar industry of Hawaii.

About ten years ago, Jesse Condé came into the Library at the Experiment Station of the Hawaiian Sugar Planters' Association and asked if he might see and use any material we had on plantation railways. Because it is the nature of the Library to have collected every scrap of information, however trivial it might at the time appear on anything germane to sugarcane and plantations, and because Miss Mabel Fraser, librarian for thirty-three years, was herself a railway buff, that request of Mr. Condé's was mutually rewarding.

Painstaking researcher that he is, Mr. Condé spent a number of years digging into those fruitful files on plantation railways. With material garnered there and from other sources to which it led him, he has already produced two remarkable books on railroading in Hawaii. The first one, *Narrow Gauge in a Kingdom*, describes common carrier pioneering in Hawaii. The second, *Sugar Trains: Narrow Gauge Rails of Hawaii*, is a chronicle of the little railways that linked the fields with the factories and the ships that carried sugar, and of the colorful men who built the little railways. These two books are the natural parents of the present volume, *Sugar Trains Pictorial*, since in the course of compiling them the photographs shown here were collected.

The other rewarding aspect of Mr. Condé's many visits to that Library was the admiration and respect generated among the staff there for his perseverance and careful attention to detail and accuracy. Together with the admiration and respect grew a warm and very real friendship with the author. All of these things together draw from us all a very sincere

ALOHA NUI, JAY!

Charlotta Hoskins
Honolulu
June 1975

Early photo, circa 1865, of Queen Emma, later destined to ride on the first steam powered cane haul railroad at Waiakea and first in the Hawaiian Kingdom. *Archives of Hawaii.*

Chapter One / Island of Hawaii

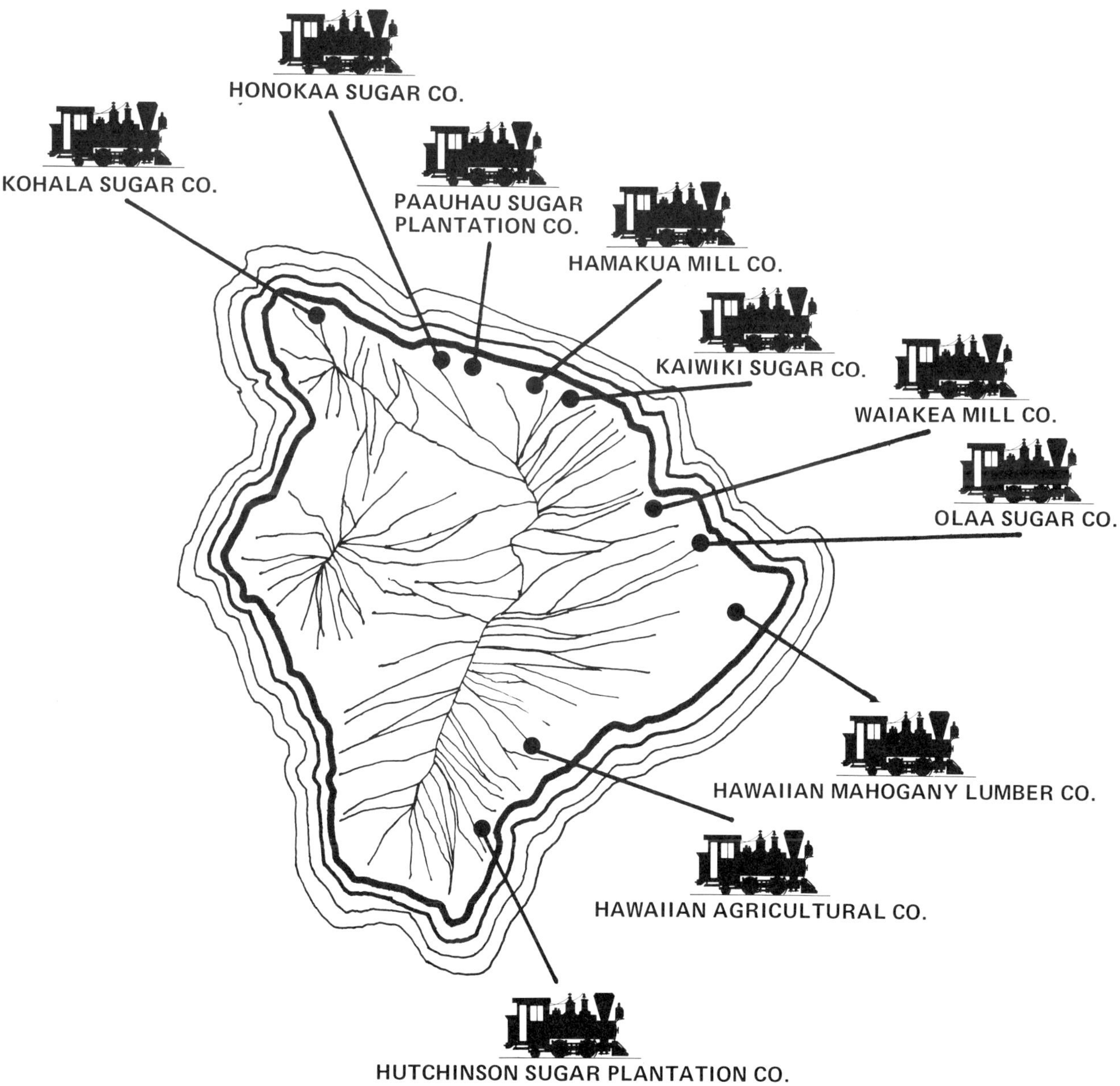

Above, H.Ag. Co. #3, 7 Ton Baldwin Switch engine — "Little Kilauea". Builder's photo 1913. Below, builder's photo of H. Ag. Co. #4, locomotive "Hiiaka" 1904. *M. H. Ferrell Collection*

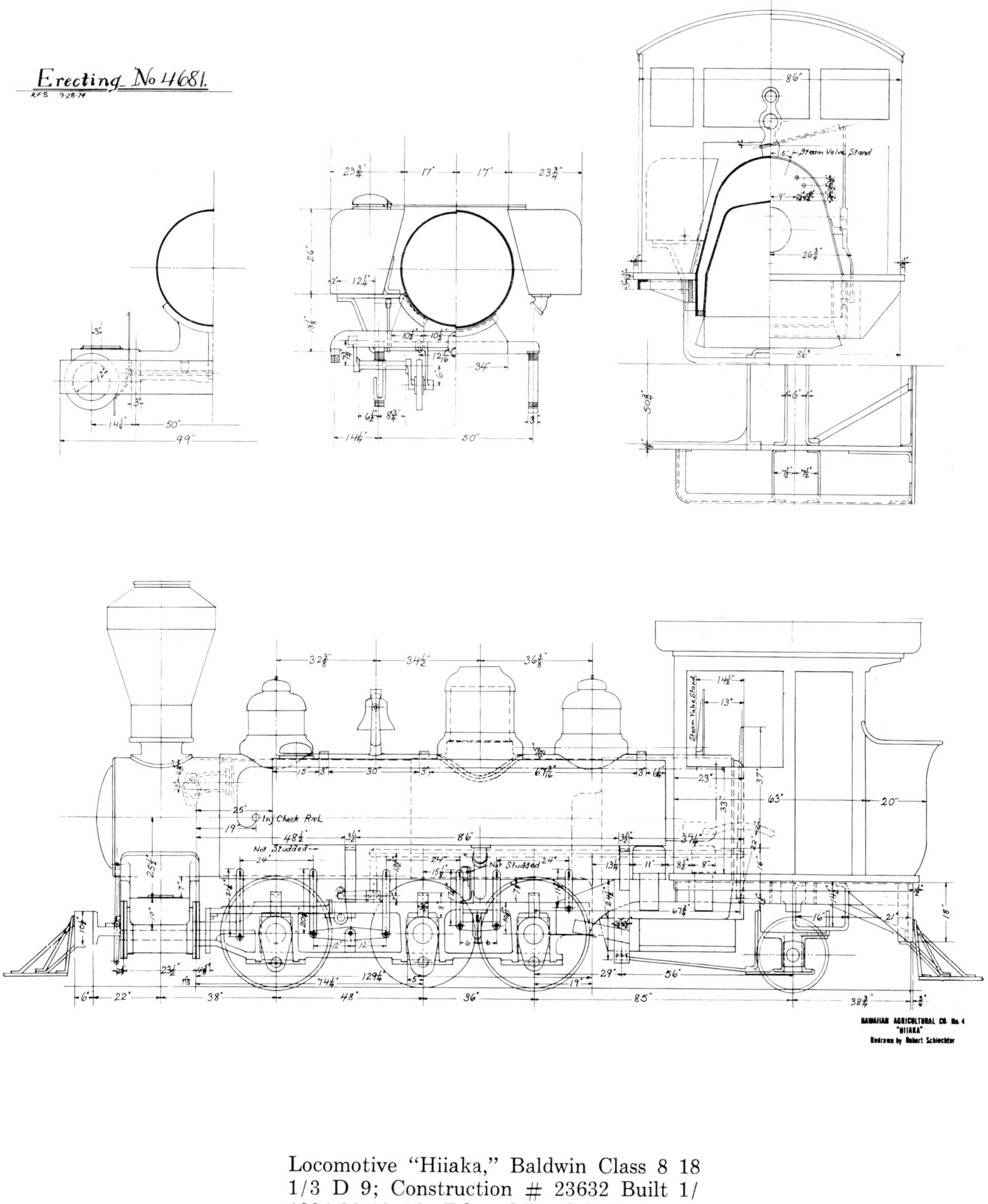

Locomotive "Hiiaka," Baldwin Class 8 18 1/3 D 9; Construction # 23632 Built 1/1904 33-12x16. *Edward Bond Collection.*

HAMAKUA MILL CO.

Loading sugar cane in February of 1925 for transport by rail to mill. One hundred tons per day. Cost: 9 cents per ton, three men and three mules. Total cost $9.00 per day. *Hawaiian Sugar Planter's Association*

HONOKAA SUGAR CO.

Heavy surf breaking on shore at Honokaa Landing about 1935. Landing derrick stays show in the left foreground. *Honokaa Sugar Company.*

Old "passenger" landing system, "Scotch Coast," Honokaa Landing, Hamakua, Hawaii, circa 1890. Freight and sugar was transferred the same way. *Archives of Hawaii*

Derrick rig, molasses line tender, and offshore vessel loading molasses at Honokaa Landing about 1935. *Honokaa Sugar Company.*

Honokaa Mill viewed from the bluffs above the mill, circa 1935. *Hawaiian Sugar Planter's Association.*

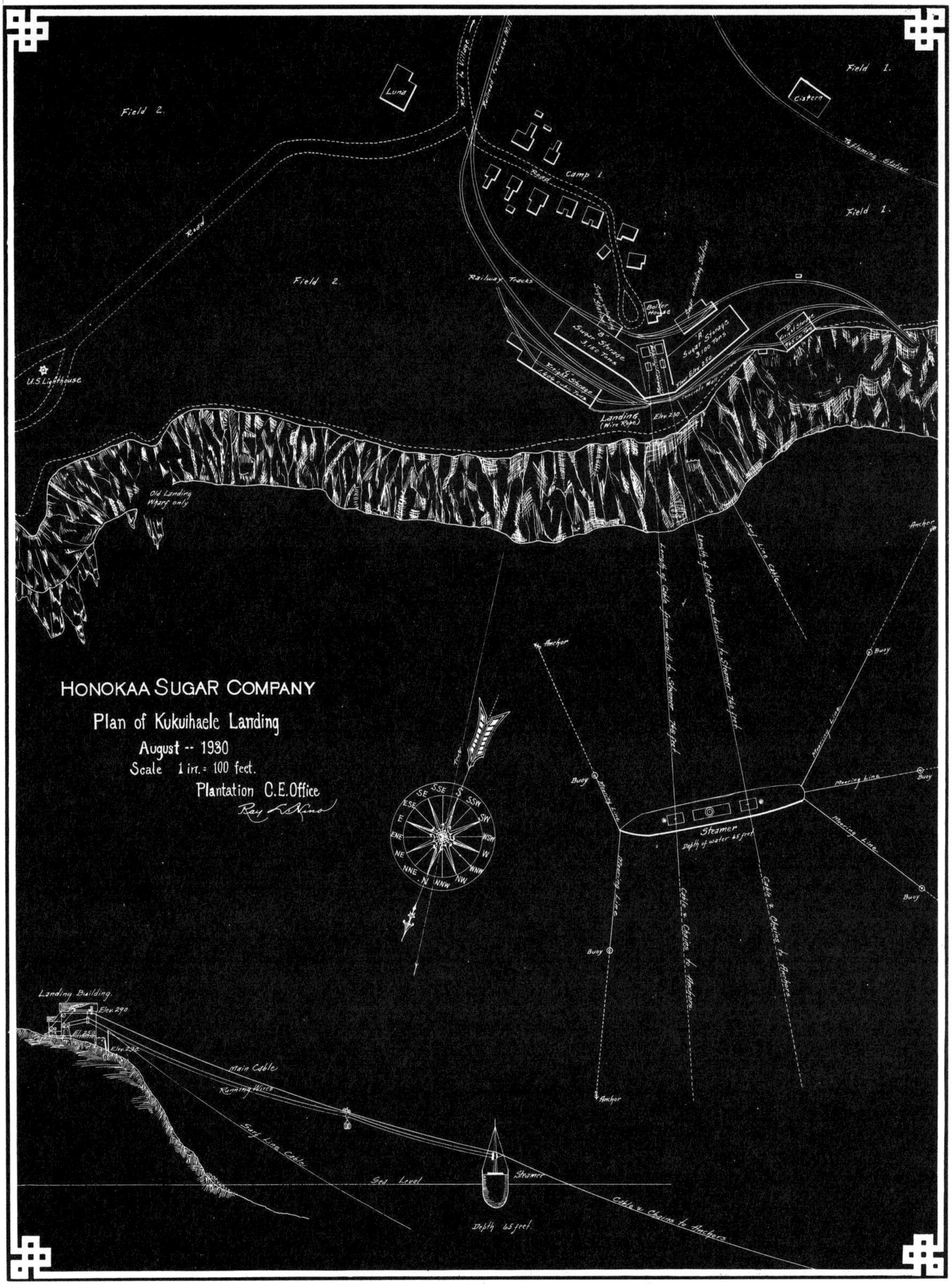

Blueprint of Kukuihaele Landing 1930. Note rail lines to hoisting and sugar warehouse building and method of anchoring off-shore vessels. *Honokaa Sugar Company.*

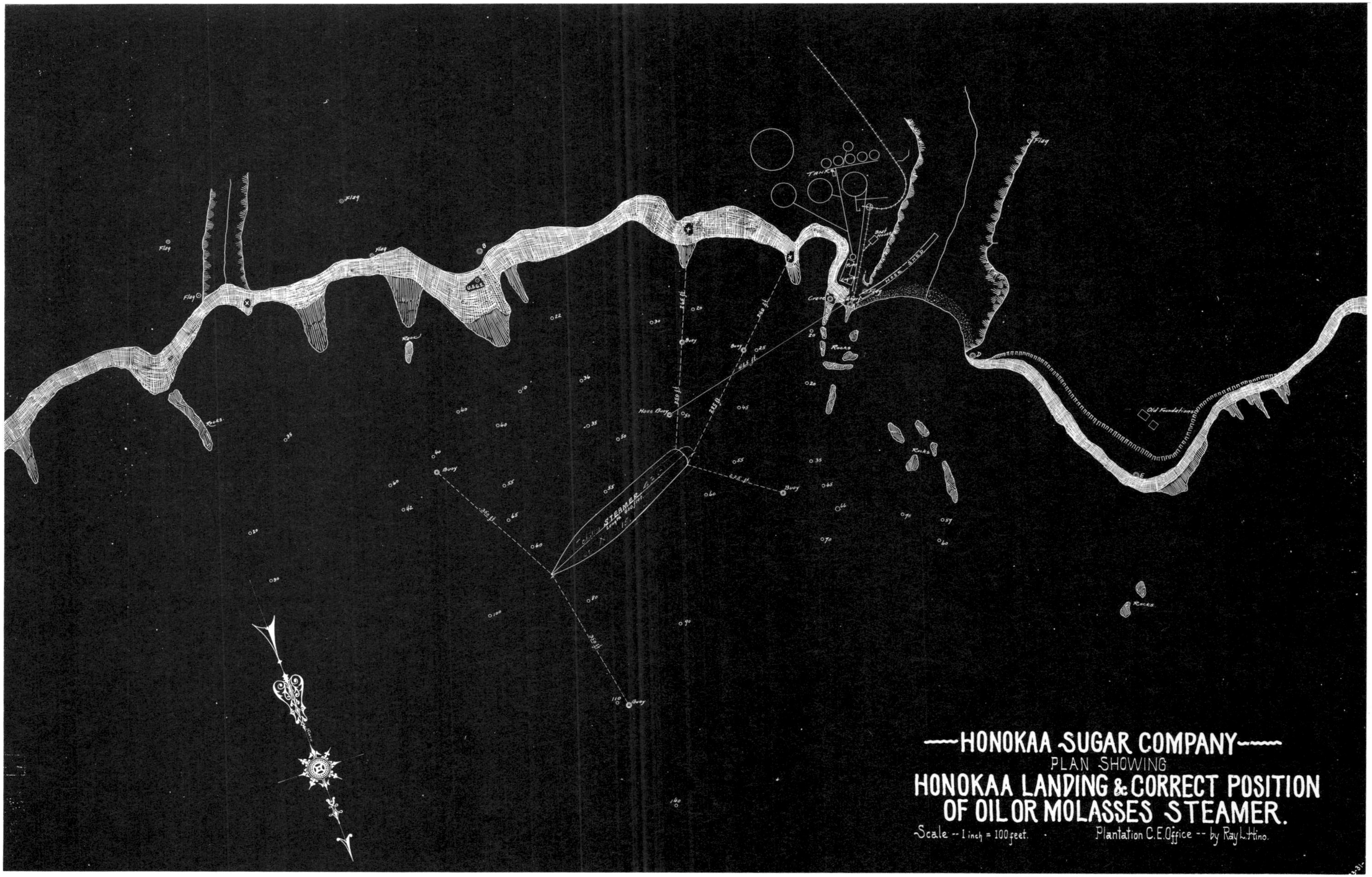

Blue print of Honokaa Landing showing molasses installation and rail connections 1931. *Honokaa Sugar Company.*

Sugar Mill of Hutchinson plantation. Date uncertain, probably about 1935. *Hawaiian Sugar Planter's Association.*

Honuapo Landing and railhead with Inter-island steamer loading sugar 1917. *Hawaiian Sugar Planter's Association.*

Later photo of Honuapo Landing, probably about the time of the dual usage by the Hawaiian Ag. Co. and Hutchinson Sugar Co. Off-shore vessel is Matson's "SS Mauna Loa." *Hawaiian Sugar Planter's Association.*

KAIWIKI SUGAR CO., LTD.

Late view of the mill at Ookala. Circa 1935.
Hawaiian Sugar Planter's Association.

KOHALA SUGAR CO.

Mill yard and storage tracks with sugar cane flume. 1938. *Bishop Museum.*

Kalakaua Coach at Mahukona 1946. The beautifully etched glass door panels of this 1882 car are said to be still in existence with a private owner. *Bishop Museum.*

Kalakaua Coach in Oahu Railway yards, Honolulu, prior to shipment to the U.S. Mainland about 1960. *Robert Kemble.*

Honuipu Landing. Sugar warehouse and cable terminal at old sugar shipping point near Puakea Point, Kohala, Hawaii. Note man riding the carrier tray. Date uncertain. *Kohala Sugar Co.*

Schooner "Muriel" being unloaded by Honuipo cable system at old landing site. Later acquired by the Kohala Sugar Co. *Kohala Sugar Co.*

OLAA SUGAR CO.

Olaa mill cane storage yard and scale house for cane cars about 1935. *Hawaiian Sugar Planter's Association.*

HAWAIIAN MAHOGANY LUMBER CO.

So-called "blind side" of the Shay locomotive used by the Lumber company at Pahoa, 1912. *Gerald M. Best Collection.*

Rear view of the lumber mill at Pahoa. Stacked railroad ties and assorted lumber in the foreground. *Paradise of the Pacific.*

Lumber yard and waste pile of Ohia Lehua at Pahoa. Circa 1910. *Paradise of the Pacific.*

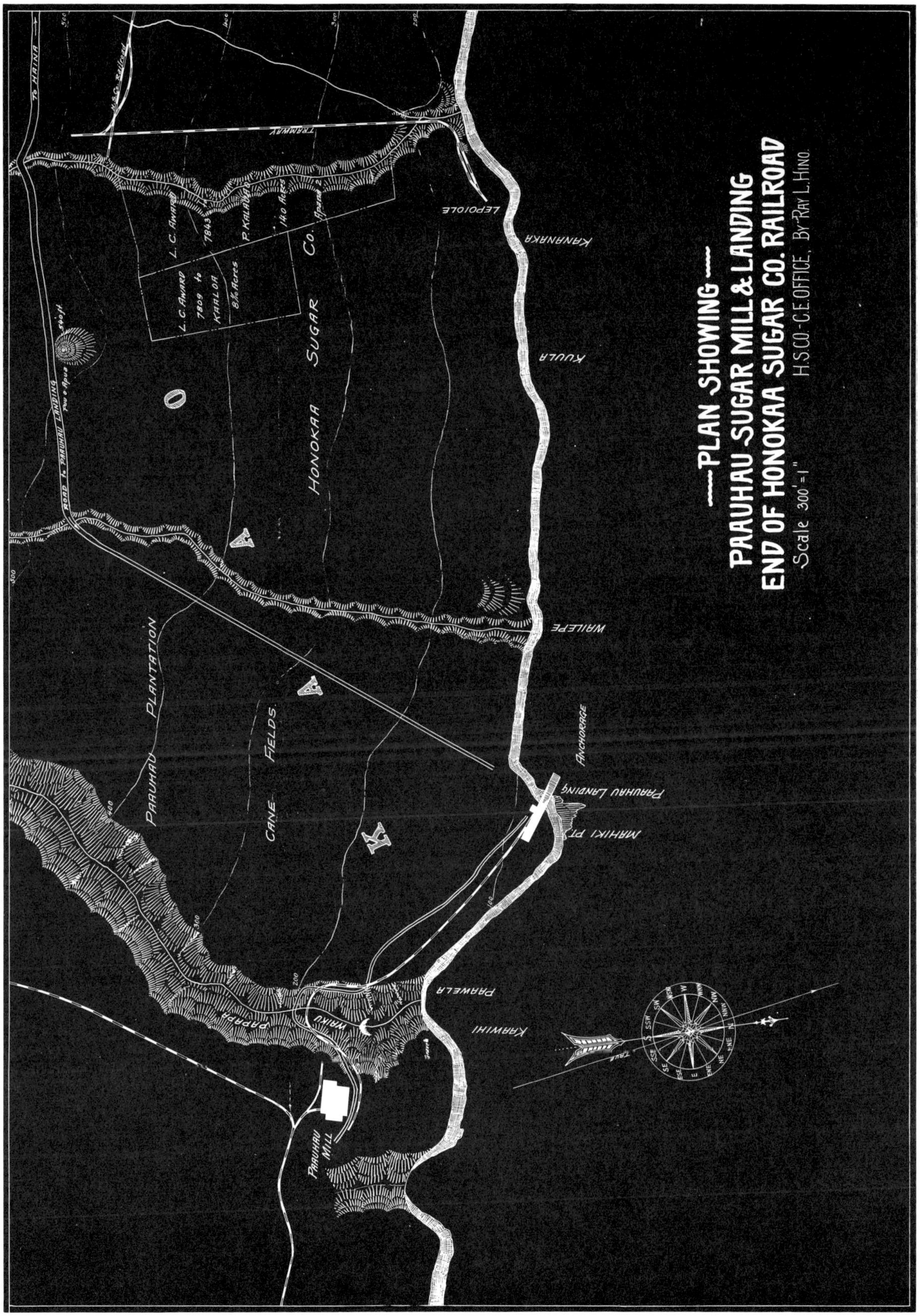

Blueprint of Paauhau Landing about 1930 showing rail connection to the landing as well as a Honokaa tramway and prominent gulches of the area. *Honokaa Sugar Co.*

Above, Waiakea locomotives "Waiolama," (left) and "Waiakea," (right) in cane field approaches to the mill yard. Date uncertain but probably about 1900. *M. H. Ferrell.* Left, First "Otto" gasolene (sic) cane hoist at Waiakea Mill. Designed and put into use by Charles C. Kennedy, plantation manager and railroad builder. Shown here near the Waiakea Roundhouse and without later cab enclosure. *C. C. Kennedy.*

Chapter Two / Island of Kauai

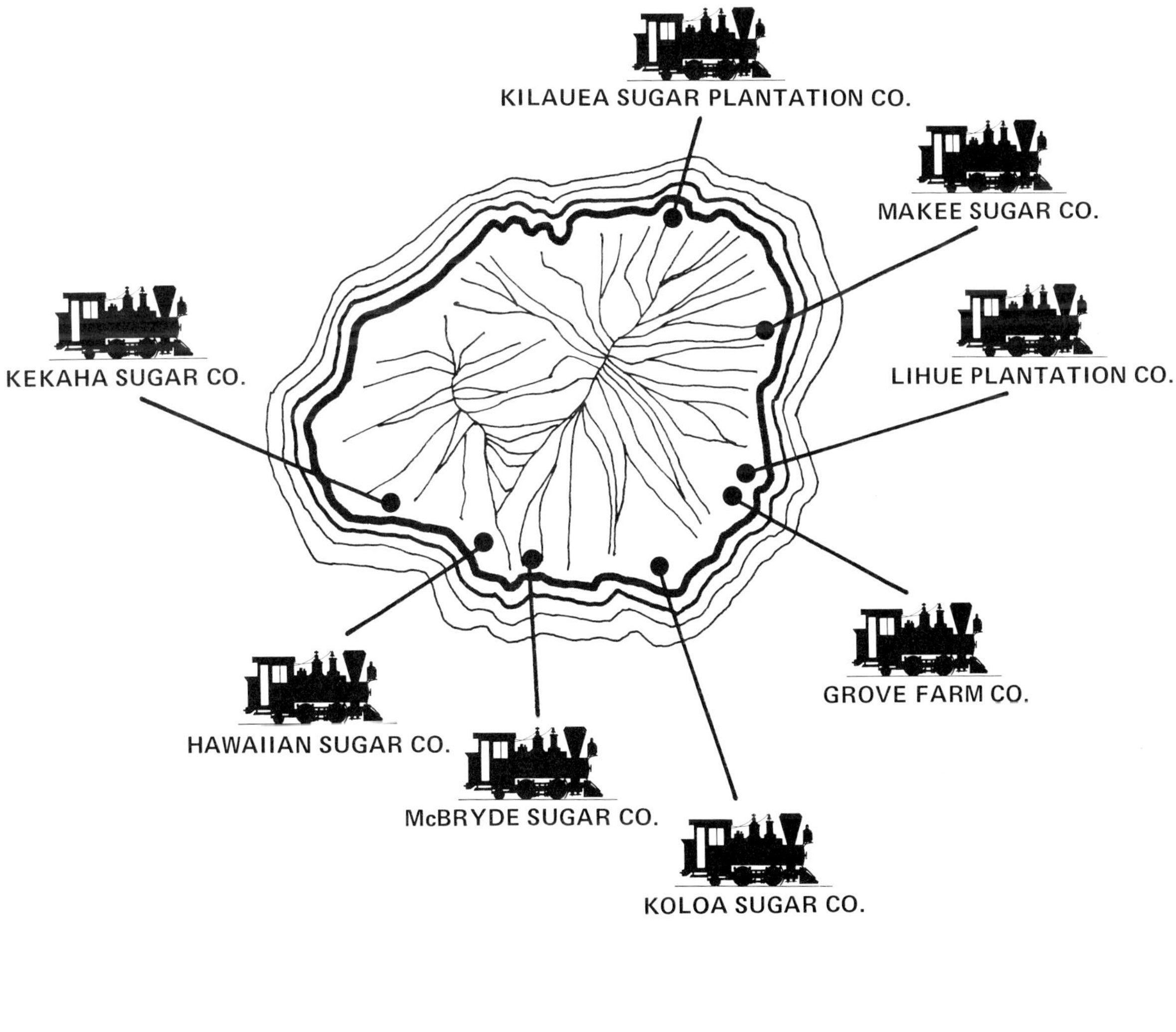

Above, Grove Farm mill with cane cars and cane cleaner about 1935. *Hawaiian Sugar Planter's Association.* Left, backhead view of 1887 locomotive "Paulo", Grove Farm carpenter shop 1972. *Photo Hawaii.*

Top, Road locomotive G.F.Co#3, "Kaipu", in retirement 1972. Center, ¾ths Front view of G.F.Co.#6, "Wainiha" at old Grove Farm roundhouse 1972. Right, Front end of G.F.Co.#5, "Wahiawa", in retirement 1972. *All: Photo Hawaii.*

Backside of H. S. Co. mill from the old Government Road about 1920. *Hawaiian Sugar Planter's Association.*

Marshalling cane cars on the portable track about 1925. *Hawaiian Sugar Planter's Association.*

KEKAHA SUGAR CO.

Kekaha Mill and cane cars entering for unloading about 1930. *Bishop Museum.*

KILAUEA SUGAR PLANTATION CO.

Chinese laborers cutting sugar cane at Kilauea about 1896. Note lunas (overseers) on horseback. *Archives of Hawaii.*

First mill and cane unloading system with "Fowler" powered sugar cane train at Kilauea plantation 1882. *Archives of Hawaii.*

Close-up of first mill at Kilauea showing detail of cane unloading system 1882. *Archives of Hawaii.*

View of Mokolea Point and Kahili Landing on Kilauea Bay, with sugar loading apparatus and Inter-island steamer at Kilauea Sugar Co. 1925. The landing went out of service in 1929. *Kilauea Sugar Co.*

New Koloa Sugar Co. mill on completion in 1913. Built by the Honolulu Iron Works and showing rail entrance to the mill. Most of the old mill equipment from Koloa Town was moved to this new site. *Grove Farm Company.*

The original Hohenzollern locomotive of 1887 as preserved at Grove Farm about 1959. *Robert Kemble.*

KOLOA SUGAR CO.

Glass plate impression of old railway cut at Koloa Plantation in the day of the "Fowler Railway Plant" about 1883. *Kauai Museum.*

MAKEE SUGAR CO.

View of old sugar mill and railroad at Kapaa. Original site of Makee Sugar Company about 1890. *Kauai Museum.*

Above, Makee stone train filling in around old wooden trestle with field stone about 1934. *Sanford Gibbs Collection, Bishop Museum.* Right, discarded locomotive boilers in the Makee Sugar Co. scrap field 1934. *Bishop Museum.* Below, Makee plantation cane train warnings as used on roads and fields of the plantation during harvesting. Note that both day and night signals were used. Circa 1932. *Hawaiian Sugar Planter's Association.*

Top, Lihue Mill and rail yards about 1900. *Grove Farm Company*. Center, later day Lihue Mill and rail yards, 1935. *Hawaiian Sugar Planter's Association*. Left, Lihue Mill interior with cane cars and unloading method. *Amfac Inc.*

Close-up view of first locomotive — "W. C. Parke" — at Lihue. From a glass plate negative of about 1895. *Kauai Museum.*

¾ rear view of the — "W. C. Parke" — with cane car. Note the neatness of the hand cut cane. From a glass plate negative exposed in the mill yard about 1895. *Bishop Museum.*

Hibiscus and Heliconia Rail Club, of Honolulu, on fan trip to Lihue 1959. Locomotive #4 was later renumbered to #604 and finally sold to a mainland buyer. *Murray Befler, Photo Hawaii.*

H & H fan trip crosses the Wailua bridge near Lihue 1959. *Robert Ramsay.*

Lihue #4 under engine servicing at Lihue in September of 1959 during H&H club railfan trip. *Robert Ramsay.*

Chapter Three / Island of Maui

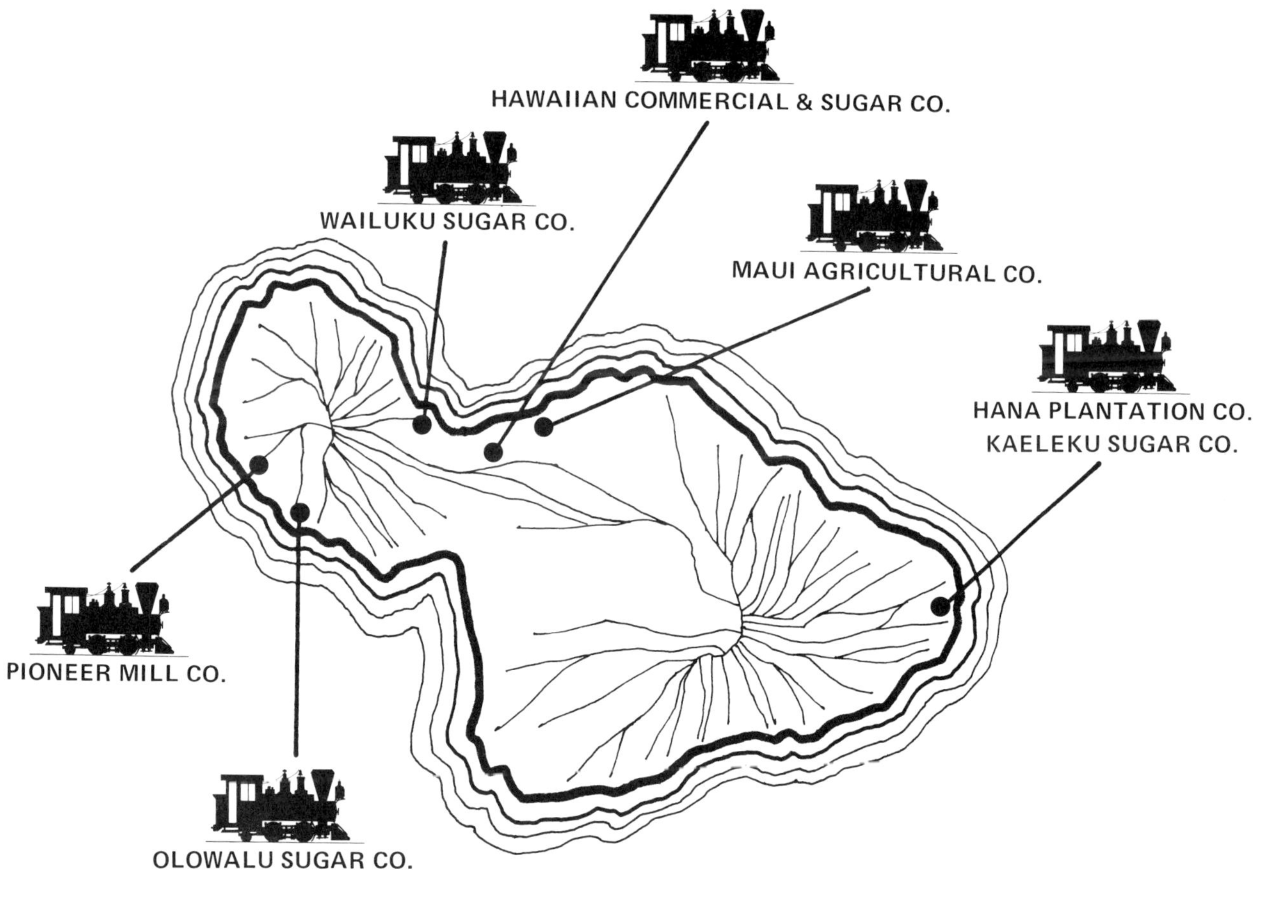

Construction of early sugar mill complex at Spreckelsville by Claus Spreckels, circa 1880. *Archives of Hawaii.*

Spreckelsville, Maui and sugar mill 1890. Cane car construction in progress (middle left) and loading of processed sugar (middle right). *Archives of Hawaii.*

New mill at Puunene with cane storage yard, circa 1905. Locomotive engine house can be seen in left middleground. *Harry Wessel Collection.*

Portable track car with mule power and crew in the Highlands of Puunene 1910. *Bishop Museum.*

Japanese women loading sugar cane cars, Hapai Ko style, Puunene, 1912. *Baker Collection, Bishop Museum.*

Experimental cane loader loading cane car at Puunene about 1915. *Hawaiian Sugar Planter's Association.*

A

Series of interior mill scenes at HC&S Co., mill about 1920, showing the machinery necessary to produce processed sugar. A. Sugar mill steam engine and driver gears. B. Sugar crushing mill roller train. C. Boiling vacuum pans. D. Sugar Evaporators. E. Sugar Crystallizers. *Baker Collection Bishop Museum.*

B

C

D

E

A. One of the first Baldwin 2 foot gauge locomotives at Spreckelsville, circa 1885. *Hawaiian Sugar Planter's Assn.* B. HC&S Road locomotive #6, H.K. Porter Co. #5026, near the enginehouse at Puunene about 1930. *Juanito Souza.* C. HC&S 3rd #3 at Puunene 1949. *Hawaiian Commercial and Sugar Co.* D. HC&S Co. 5th #4 at Puunene 1949. *Hawaiian Commercial and Sugar Co.* E. HC&S Co. 2nd #5 at Puunene 1949. *Hawaiian Commercial and Sugar Co.*

A

B

C

D

E

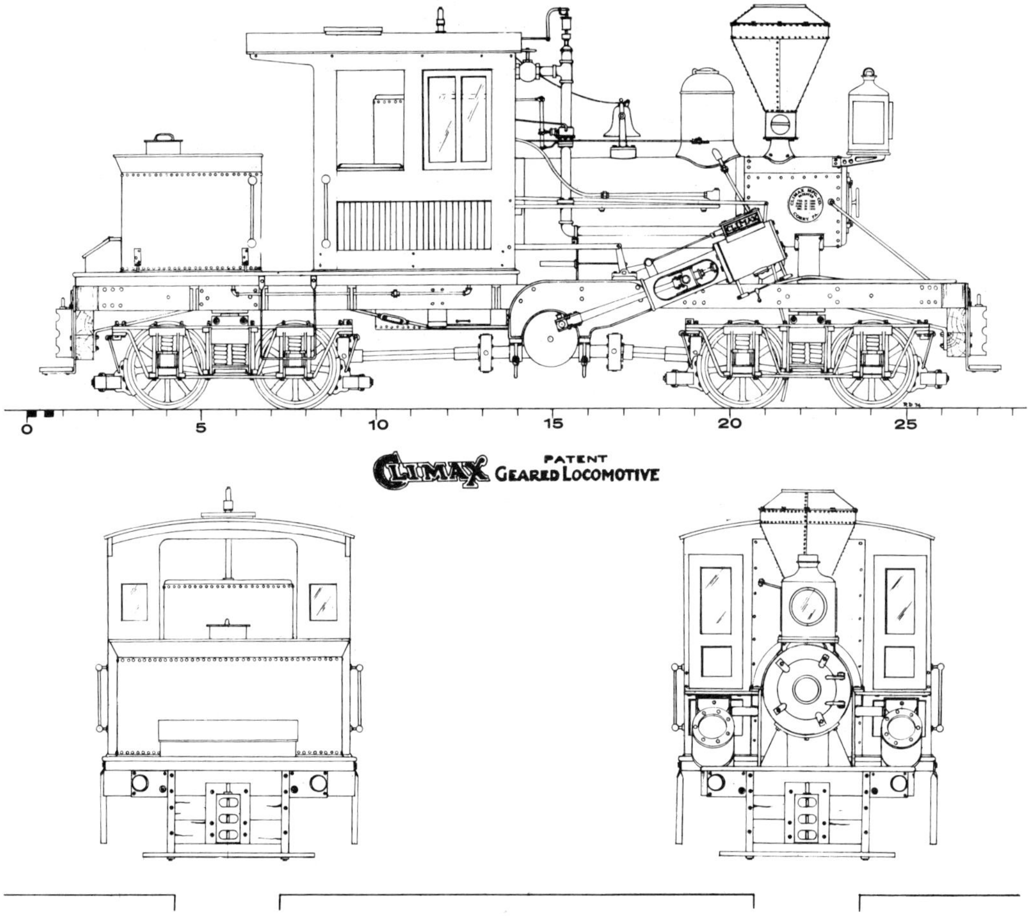

18 TON CLIMAX LOCOMOTIVE CIRCA 1900, 28" DRIVERS, $8\frac{1}{2}$" X 12" CYLINDERS

Experimental Climax, circa 1906. Believed to be the only locomotive of this style ever sent to Hawaii. *Rich Dunn drawing.*

MAUI AGRICULTURAL CO.

Sugar mill and railroad crossing sign — "Alanui Hao" (Great Iron Path) at Paia prior to 1918. *Baker Collection, Bishop Museum.*

Northwest cane loader loading cane car train at Maui Ag. Co. 1927. *Hawaiian Commercial and Sugar Company.*

M.Ag. Co. #5 "Pulehu" in plantation service 1934 and just after the locomotive had released a string of cars to an adjacent storage track. The tender has been relettered since it came from the factory in 1916 and the cab window sports a brass name plate in place of the original painted name. *Henry Bonnell.*

Maui Ag. Co. #2, "Maui" in the erecting shop Baldwin Locomotive Works 1904. *Louis Saillard.*

Unloading cut cane from rail cars at Hana Plantation about 1905. *Ovenden Collection, Bishop Museum.*

Loading processed sugar onto schooner (masts can be seen through the wharf) at Hana Bay about 1905. *Ovenden Collection, Bishop Museum.*

Lightering processed sugar to steamer in Hana Bay from railroad wharf of the Kaeleku Sugar Co., circa 1925. *Hawaiian Sugar Planter's Association.*

Pioneer Mill Company office. Date uncertain. *Baker Collection, Bishop Museum.*

Railyard and mill at the P.M. Co. about 1920. Mules in use to move rail cars to mill unloader. *Amfac Inc.*

Close-up view of cane car unloader at Pioneer Mill about 1925. *Baker Collection, Bishop Museum.*

The Kaanapali Landing of Pioneer Mill Co. Off-loading sugar bags to an American Hawaiian steamer from the rail head, 1912. *Baker Collection, Bishop Museum.*

OLOWALU SUGAR CO.

5 Ton diesel locomotive and train of empty cars at Olowalu Plantation, a two-foot gauge train on the main line about 1936. *Hawaiian Sugar Planter's Association.*

WAILUKU SUGAR CO.

Wailuku mill with hand cut cane on 30 inch gauge cars entering the mill about 1930. *Archives of Hawaii.*

Chapter Four / Island of Molokai

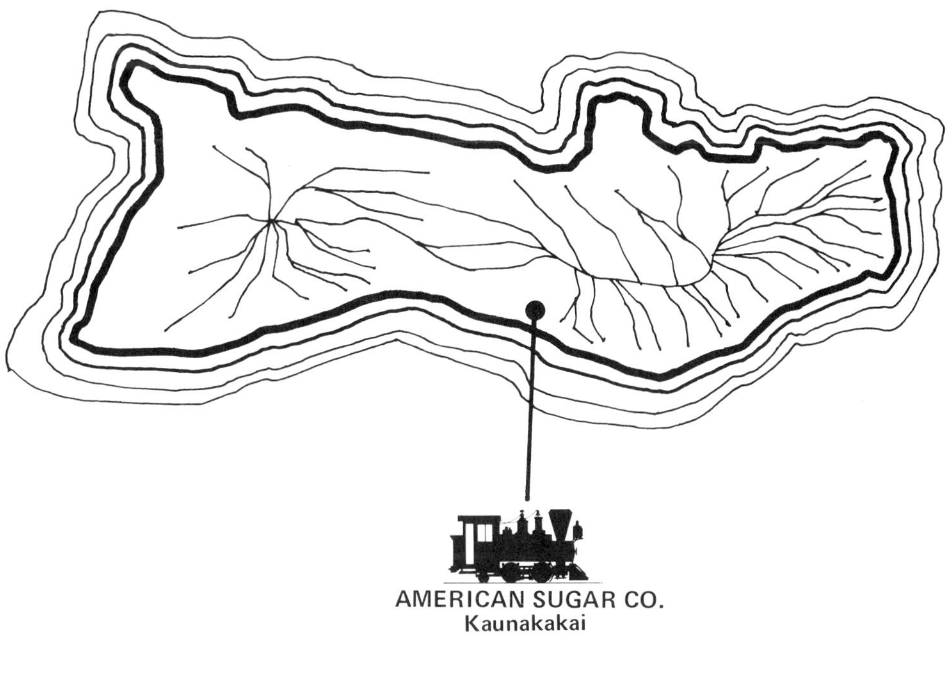

Above, Road locomotive "Uncle Sam" of ASCO 1899. The gentleman is Charles M. Kurtz, author of "Track & Turnout Engineering" and builder of the railroad at Kaunakakai. The locomotive is a Baldwin 0-6-2TB, shown before its sale to the Honolulu Sugar Co. at which time it was altered to a "tank over boiler" machine about 1901. *Eric Thomsen Collection.* Below, As far as could be found out, this gentle lady is Mrs. Kurtz. Standing in the gangway of the "Uncle Sam" in 1899 she seems to say — "How proud I am of my husband, the Railroad Builder." *Eric Thomsen Collection.*

Chapter Five / Island of Oahu

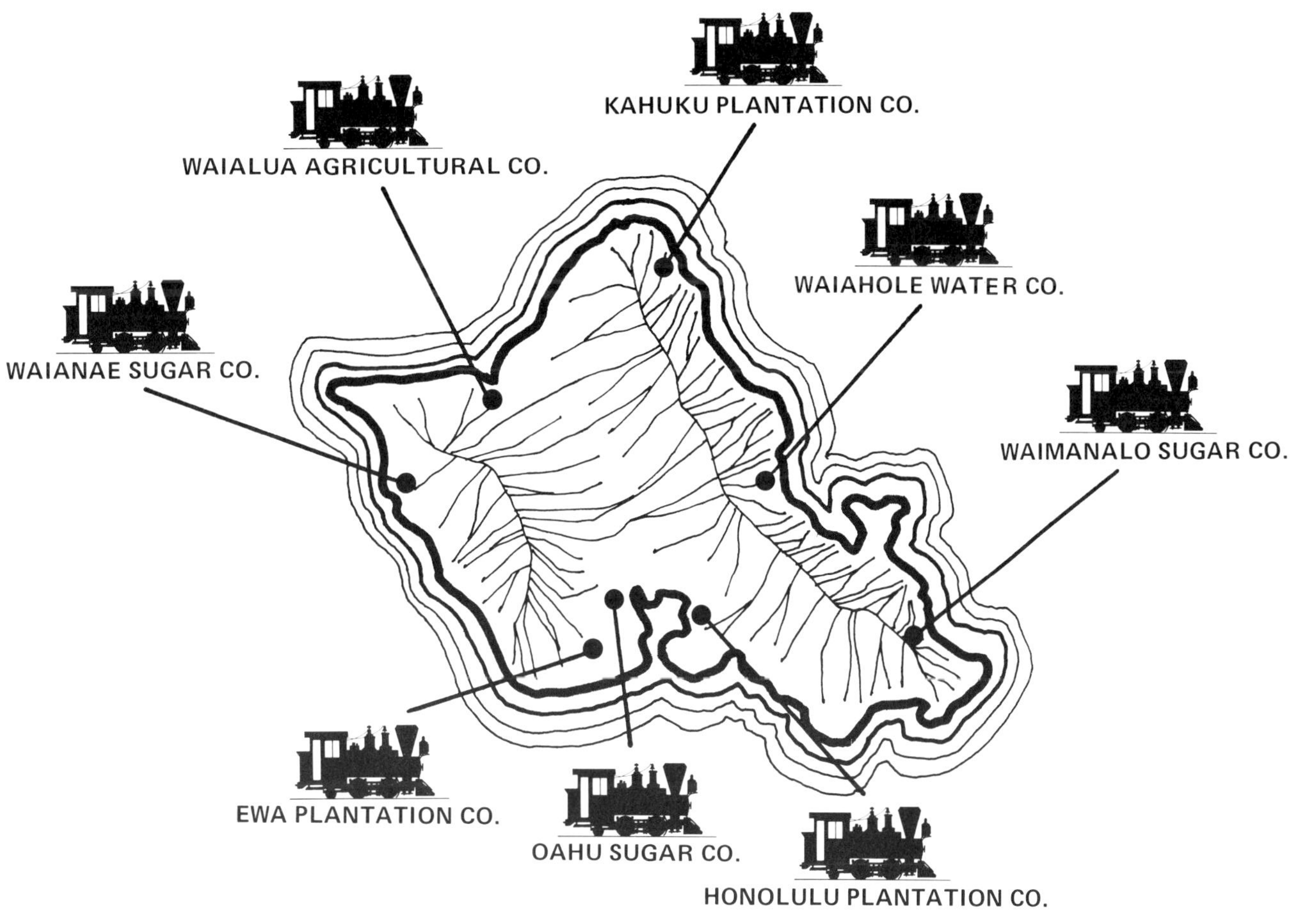

Early view of Ewa mill with cane cars moving up to supply two crushing mills. Date uncertain, probably about 1905. *Archives of Hawaii.*

Ewa roundhouse and turntable. Built in 1924 the roundhouse and "armstrong" (manually operated) turntable were still in use in 1946. *Cochrane Collection, Bishop Museum.*

EWA PLANTATION CO.

E.P. Co. #1, "Ewa", in plantation service April 5, 1923. *Richard B. Jackson Collection.*

E.P. Co., "Ewa" hauling derrick style cane loader near the mill in 1937. *Mallory Hope Ferrell Collection.*

The "Sampson" runs behind the old Ewa roundhouse for R. B. Jackson on May 1, 1923 near the Ewa mill. *R. B. Jackson Collection.*

Ewa Plantation Co. #5, "Nanakuli" in field service at Ewa May 1, 1923. *R. B. Jackson Collection.*

E.P. Co. #6 with a long cane train in the 1940's. *M. H. Ferrell Collection.*

E.P.Co. #7, the "George F. Renton Jr." on portable track standby in the 1920's. *M. H. Ferrell Collection.*

Above, E.P. Co., #4, "Sampson", on field standby close to an irrigation ditch, October 1947. *Gerald Bolton, Honolulu.* Right, E.P. Co. #6 as a last surviving cane haul locomotive on a high trestle at Ewa. This photo furnished the cover for the plantation annual report for 1947. Local reporting noted that just before photo time the locomotive broke down so it and the cane train were positioned by the plantation diesel. (E.P.Co. #8). *Photo Hawaii.*

Road locomotives "Sampson," Nanakuli" and E.P.Co. #6 triple-head in the Ewa fields in 1923. *R. B. Jackson Collection.*

Bagging processed sugar for export at Ewa. Date uncertain. *Bishop Museum.*

Kahuku Plantation #1 in reserve in mill yard 1945. *Fred Stindt G. M. Best Collection.*

K.P.Co. #1, "Keana" switching OR&L Co. cars in Kahuku yards 1946. *Cochrane Collection, Bishop Museum.*

K.P.Co. #4 inbound to the mill with field machinery and loaded cane train at Kahuku 1946. *Cochrane Collection, Bishop Museum.*

Sunday morning at the Kahuku yards 1950. Engine house shows in the background and "rip track" to the left. The "watchman" and his helper stare at the camera. *Robert Ramsay.*

Ex Navy Diesel in the Kahuku mill yards as K.P.Co. 2nd #1 in 1950. *R. Ramsay.*

KAHUKU PLANTATION CO.

Fricke chain side cane cars in the plantation storage yards at Kahuku 1950. Note extra slatting added to original car ends to allow the stowage of more cane stalks in the cars. *R. Ramsay.*

K.P.Co. 3rd #2. Whitcomb diesel on the point for a Hibiscus & Heliconia rail fan trip to Hauula about 1950. *R. Ramsay.*

Oahu Railway & Land Company passenger cars at the old Kahuku Station in the service of the Heliconia & Hibiscus Short Line Railroad about 1950. *R. Ramsay.*

View of Oahu Sugar Co. mill with cane cars in storage and cars with hand cut cane moving in for crushing. Circa 1900. *Archives of Hawaii.*

R. B. Jackson scene of rail activities at Oahu Sugar Co. May 1923. *R. B. Jackson Collection.*

Sugar crushing mill machinery in motion at Oahu Sugar Co. Circa 1910. *Archives of Hawaii.*

O.S. Co. #1, "Waipahu" switching tank cars in the mill yard 1946. *Cochrane Collection, Bishop Museum.*

Oahu Sugar Co. #1. Ex "Waipahu" poses with Dale Gentry, her owner, at the throttle, under the banner of the "Cottonwood & Southern" at Mr. Gentry's "Old Woman Ranch" about 1965. *Jack Rutherglen Collection.*

Left side view of O.S.Co. #2, "Waipio" on standby in the mill yard May 31, 1923. *R. B. Jackson Collection.*

Right side view of "Waipio" making up a train of empty cane cars for return to the fields at Waipahu May 1923. *R. B. Jackson Collection.*

O.S.Co. #3, "Waikele" switching cane cars in the mill yard at Waipahu about 1936. *Baker and Wolters, Oahu Sugar Co.*

O.S.Co. #3 hauling empty cane cars to the fields about 1946. *Cochrane Collection, Bishop Museum.*

"Waikakalaua," O.S.Co. #5 and crew await new orders at Waipahu May 31, 1923. *R. B. Jackson Collection.*

O.S.Co. #6, "Koalipea" moves up the service track in the mill yard in 1946. *Cochrane Collection, Bishop Museum.*

O.S.Co. #4, "Waiawa" on the mill yard standby track 1946. *Cochrane Collection, Bishop Museum.*

"Waiawa" moves off to a new assignment from the mill yard at Waipahu on May 31, 1923. *R. B. Jackson Collection.*

"Puuloa," O.S. Co. 2nd #7, moves ex Kahului Railroad flats, made into heavy tonnage cars, in the mill yards about 1950. *H.S.P.A.*

H.K. Porter locomotive, O.S.Co. #8, "Hoaeae" backs down to the mill with loaded cane cars over a low trestle near the mill in 1946. *Cochrane Collection.*

"Hoaeae" takes on oil in the Oahu Sugar Co. service area at Waipahu 1946. *Cochrane Collection, Bishop Museum.*

"Waikane," O.S.Co. #9 heads home with work gangs backing down to the mill. Note women dressed in typical cane cutter's garb in first car. *Baker and Wolters Oahu Sugar Co*

Oahu Sugar Co. locomotives #3 and #6 at World War II plantation checkpoint. An Anti-aircraft Battery in 1943. *H.S.P.A.*

Early scene at Oahu Sugar Company, cutting and stacking sugar cane. Cane cars and mules appear in the scene. Circa 1899. *Baker and Wolters, Oahu Sugar Co.*

Marshalling cars of cut cane with mules on the portable track at Oahu Sugar Co. Circa 1900. *H.S.P.A.*

Grab loading method of filling rail cars with cut cane at the Oahu Sugar Co. in 1936. *Baker and Wolters Oahu Sugar Co.*

Emptying "mud press", (Juice Sediment) rail cars at Oahu Sugar Co. in 1936. This sediment was often used on the cane fields for fertilizer. *Baker and Wolters Oahu Sugar Co.*

"Lining out" or spotting empty cane cars on portable track prior to loading at Oahu Sugar Co. Date uncertain. These are Fricke chain-side cars. *Amfac Inc.*

O.S. Co. locomotives double heading a long cane train from the sugar fields near Waipahu. Date uncertain. *Archives of Hawaii.*

Honolulu Plantation mill with O.R.&L.Co. cars loading sugar under the mill portico about 1915. *Hawaiian Sugar Planter's Association.*

Road locomotive "Halawa" in H.P.Co. plantation service before engine numbers were applied. Circa 1906. *Fred Denison Collection H.S.P.A.*

Oahu Railway locomotive #32 switching box cars at Honolulu plantation 1946. *Cochrane Collection, Bishop Museum.*

Bagging processed sugar at H.P. Co. Date uncertain. *Archives of Hawaii.*

The camera of R. B. Jackson captures "Waimalu," H.P.Co. #3 in cane field service near Aiea May 31, 1923. *R. B. Jackson Collection.*

WAIAHOLE WATER CO.

South Portal track of the water company about 1914. A Shay locomotive can be seen in the middle background. Note the size of the boulders pushed aside for laying track. *John Knaus.*

Early photo of Waialua mill of about 1900. Far locomotive is a Baldwin machine while nearer engine is a Fowler locomotive. Basic cane storage yard of the times with cane cars being pushed into the mill. *C. C. Kennedy.*

Aerial view of Waialua Ag. Co. mill about 1935. *H.S.P.A.*

View of the interior of the plantation machine shop at Waialua about 1935. Repairs for locomotives and all plantation equipment were made here. *H.S.P.A.*

Side view of W.A.Co. #1 in retirement at Waialua 1946. *Cochrane Collection, Bishop Museum.*

W.A.Co. 2nd #2, Ex Kahului Rr. Co. #5 in the yards 1946. *Cochrane Collection, Bishop Museum.*

W.A.Co. #4, outbound to fields with empty cane cars and the usual sand car. *Author's Collection.*

W.A.Co. #5 on the "point" during employee rides at Waialua 1952. *Author's Collection.*

W.A.Co. #6 in plantation service with long cane train. Date uncertain. *H.S.P.A.*

W.A.Co. #7 backing down to the mill with a cane train about 1947. *Photo Hawaii.*

W.A.Co. rip track with locomotives lined up for scrapping about 1952. *Author's Collection.*

W.A.Co. #3 believed to be enroute for scrapping about 1952. *Author's Collection.*

Left side view of "Huki Moku" a Fowler locomotive in the O.R.&L.Co. scrap yard. Date uncertain but probably about 1914. This is a similar style locomotive as appears in the Waialua mill yard photo. *Fred Denison Collection H.S.P.A.*

Cane cars at the mill, "Full in — Empty Out". Date uncertain. *H.S.P.A.*

W.A.Co. #4 on standby with sand car in plantation yards. Date uncertain. *Author's Collection.*

W.A.Co. #6 Home built locomotive on the "point" with loaded cane cars for the mill. Date uncertain. *Author's Collection.*

W.A.Co. #7 Ex Kahului Rr. #8 in retirement at Waialua Mill 1947. *Cochrane Collection, Bishop Museum.*

W.A.Co. 2nd #2 switching cars on the plantation line near the Oahu Railway main line 1946. *Cochrane Collection, Bishop Museum.*

W.A.C. #6 as partially restored to steam operation at Laulualei, Naval Ammunition Depot, Oahu, June 18, 1972. *John Knaus.*

¾ Front view of W.A.Co. #8, Plymouth yard switcher pulling the inevitable sand car and supplies. Date uncertain. *John Knaus.*

Side view of W.A.Co. #8 at Waialua 1950. *Robert Ramsay.*

View of W.A.Co. #11 a diesel unit in the mill yard about 1951. *Robert Ramsay.*

The engine house at Waialua. W.A.Co. #5 is on standby at the switch. The track at the far left was the plantation connection with the OR&L.Co. main line. *Robert Ramsay.*

A Weed-killer tank car at Waialua Ag. Co. Home made by the plantation machine shop, in the yards about 1950. *Robert Ramsay.*

Early view of Waianae mill buildings, circa 1881. The square smokestack seen here is reported to have been replaced with a larger round one when bigger boilers were installed in the mill about 1884. *Archives of Hawaii.*

View of the later day mill at Waianae with Chinese unloading cane cars onto the mill conveyer about 1885. *Archives of Hawaii.*

Original locomotive "Waianae" in the plantation mill yards about 1906. A Ransomes & Rapier product it is shepherding a cane car of the times on the mill stand-by track. *Fred Denison Collection, H.S.P.A.*

The "Fowler" locomotive "Kaala" of the Waianae Plantation in the high cane of Makaha Section fields. *Archives of Hawaii.*

W.S.Co. "Kaala" at plantation service facilities as caught by the camera of R. B. Jackson, May 4, 1923. *R. B. Jackson Collection.*

W.S.Co. 2nd "Waianae" at plantation service facility April 4, 1923. *R. B. Jackson Collection.*

2nd "Waianae" as W.S.Co. #2 in retirement at the old enginehouse at Waianae 1946. *Vitaly Uzoff, Doug Richter Collection.*

W.S.Co. "H. A. Widemann" by the watertower in 1923. *R. B. Jackson Collection.*

Waianae plantation #1, "H. A. Widemann" entering the mill yard with a loaded cane train 1946. Even with side tanks and a homemade tender #1 was hard-put to bring in the cane from far-off fields without extra servicing. *Cochrane Collection, Bishop Museum.*

View of early mill arrangement at Waimanalo. Date uncertain. Note round top box cars for sugar carriage and storage, near water tank. *Bishop Museum.*

Latter day mill at Waimanalo probably about 1900. Blur near the center of the photo is a locomotive moving cane cars into the mill. While the camera stood still the engine didn't. *Archives of Hawaii.*

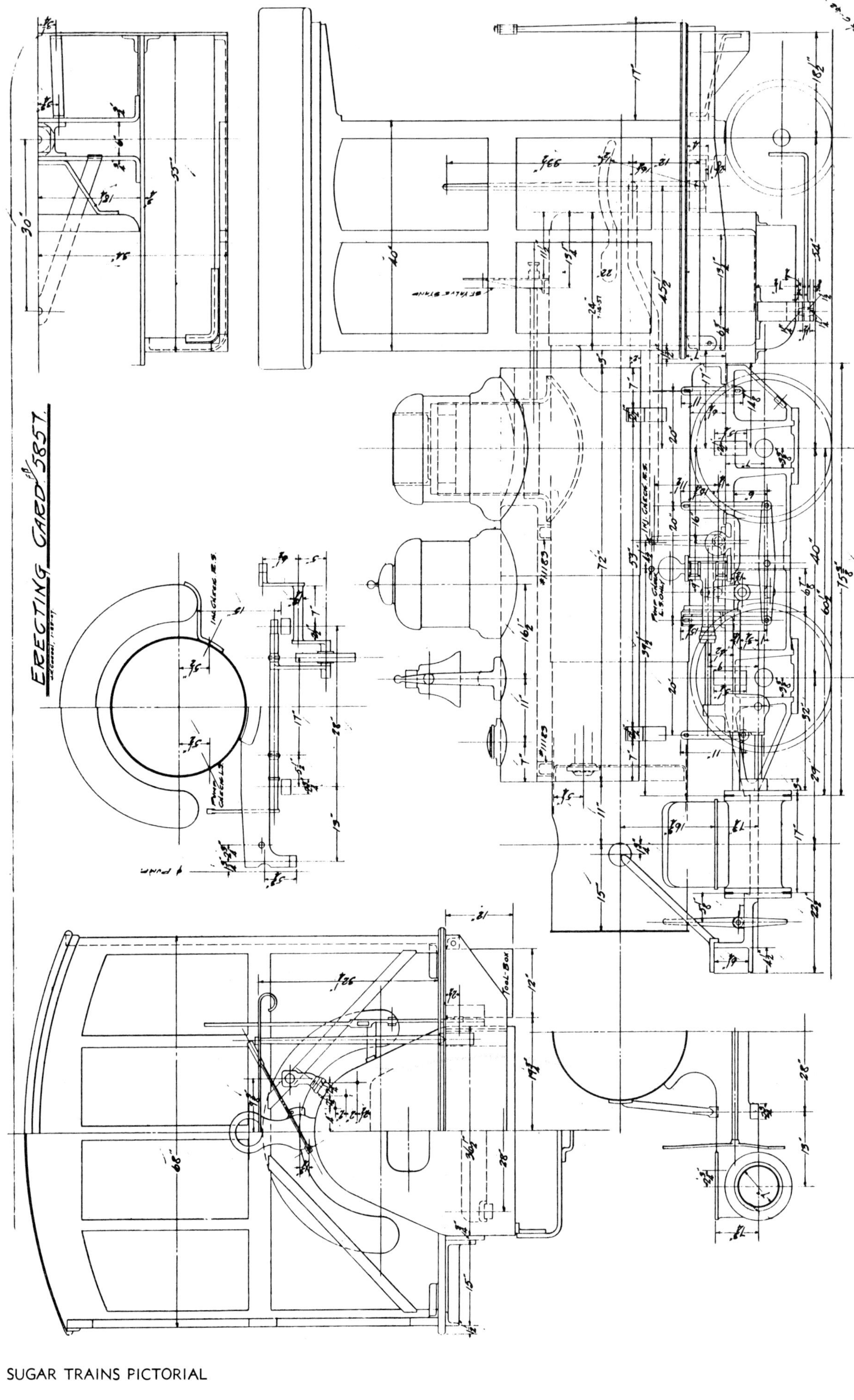

Locomotive "Pokaa. Baldwin Class 6 8 1/3
C 62; Construction # 30340. Built 2/1907
24-7x10. *Edward Bond Collection.*

Night work was in vogue at Waimanalo plantation in 1930. Note dynamo and headlight additions to "Pokaa." *Lane Sharman.*

"Nalo" heads out for work in May 1923 at Waimanalo. *R. B. Jackson Collection.*

"Olomana" in plantation service at W.S.Co. with a loaded cane train May 25, 1923. *R. B. Jackson.*

"Olomana" blasts cane cars towards the new mill at Waimanalo in May of 1923. *R. B. Jackson.*

Typical cane field portable track lay-out at Waimanalo in 1935. On this uncertain foundation tons of sugar cane were moved to the mill. *H.S.P.A.*

The camera of William Twiggs-Smith catches a mill bound cane train at Waimanalo plantation about 1935. *H.S.P.A.*

Hapai Ko under the gaze of the Olomana peaks at Waimanalo plantation 1935. *H.S.P.A.*

Chapter Six / Gregg Cane Cars

ALL AGREEMENTS SUBJECT TO DELAYS OF CARRIERS, FIRES, STRIKES OR OTHER CONTINGENCIES BEYOND OUR CONTROL. PRICES SUBJECT TO CHANGE WITHOUT NOTICE.

WM. C. GREGG, PREST. D. W. SNEDDEN, VICE PREST. C. E. STAFFORD, TREAS. J. L. DAHL, SEC'Y. L. D. GREGG, DIRECTOR.

THE GREGG COMPANY, Ltd.

MANUFACTURERS AND EXPORTERS OF

PLANTATION CARS AND RAILWAY EQUIPMENT, STANDARD AND NARROW GAUGE,
PORTABLE TRACK, FROGS, SWITCHES, CAR PARTS, FORGINGS,
CAR WHEELS, CANE HOISTS AND UNLOADERS.

MAIN OFFICE AND WORKS:
HACKENSACK, N.J., U.S.A.

BRANCH OFFICES:
NEW YORK CITY. 80 WALL ST.
HONOLULU, HAWAII, 726-744 ALAKEA ST.
SAN JUAN, PORTO RICO, 18 TETUAN ST.
HAVANA, CUBA, LONJA DEL COMERCIO, 508
LONDON, E.C. ENGLAND, 139 CANNON ST.

CABLE ADDRESS, ALL OFFICES:
"CARGREGG"

CODES USED:
A.B.C. 5TH EDITION
WESTERN UNION
EXCELSIOR CODE SYSTEM
ENGINEERING CODE
LIEBER'S
AND
PRIVATE CODE IN CATALOGUE

HONOLULU, T.H. August 19th, 1914

Lihue Plantation Co.,

Mr. F. Weber, Mgr.,

Lihue, Kauai

Gentlemen:-

SUGAR IS SELLING AT 6.52¢, OR $130.40 PER TON TO-DAY.

High prices for sugar due to the effects of the present European war will last for some time to come. Food stuffs have already risen in price. The prices of all raw materials will probably soon be effected. This includes iron and steel.

In view of the foregoing is it not obvious that now is the time to attend to your present and future needs in the line of railway equipment, track, unloader spares, plow rope, whiffletrees, tank irons, etc?

Your overhead expenses in your mill, per ton of sugar, will be reduced if your railway line is completely equipped, thus avoiding costly stoppages.

A copy of our Honolulu stock list showing railway material now on hand has recently been mailed to you.

Order now and reduce your expenses.

CBG/G #14

Yours truly,

WM. C. GREGG, PRESIDENT.
J. A. TUTHILL, VICE PRESIDENT.

C. E. STAFFORD, SECRETARY AND TREASURER.

DIRECTORS: H. W. SEAGER, D. W. SNEDDEN. J. L. DAHL, C. M. LOVSTED.

THE GREGG COMPANY, Ltd.

MANUFACTURERS AND EXPORTERS OF

PLANTATION RAILWAY EQUIPMENT
AND FIELD APPLIANCES.

"GREGG PATENT" SUGAR CANE CARS, PORTABLE TRACK,
CANE CAR UNLOADERS, ETC., STEAM PLOW
PARTS AND WIRE CABLE.

INDUSTRIAL RAILWAY EQUIPMENT.

FABRICANTES Y EXPORTADORES DE

ACCESORIOS PARA CAMPO Y EQUIPOS DE
FERROCARRILES PARA INGENIOS.

VAGONETAS PARA CAÑA DE AZÚCAR, VÍAS PORTÁTILES,
DESCARGADORES DE CARROS PARA CAÑA, ETC,
"PATENTE GREGG," CABLES DE ALAMBRE
Y PIEZAS PARA ARADOS DE VAPOR.

EQUIPOS PARA FERROCARRIL INDUSTRIAL.

MAIN OFFICE AND WORKS — OFICINA CENTRAL Y TALLERES

NEWBURGH, N. Y., U. S. A.

CABLE ADDRESS: "GREGG." NEWBURGH.

DIRECCIÓN TELEGRÁFICA: "GREGG," NEWBURGH.

BRANCHES:

Honolulu, Hawaii; San Juan, Porto Rico; Havana, Cuba; Mexico City, Mex.

SUCURSALES:

Honolulu, Hawaii; San Juan de Puerto Rico; Habana, Cuba; México México.

CATALOGUE No. 32. 1903.

PLANTERS of the Hawaiian Islands are acknowledged leaders in economical production of sugar. They have adopted the portable railroad system as the best and cheapest means of transporting the cane from field to mill. We have invented several improvements for use in this system, and claim for our "Gregg Patent" portable track and cane-cars many points of superiority. A careful examination of the following pages will show you what we have to offer and we believe you will quickly recognize the advantages of the "Gregg" system over any other means of conveying cane to the mill.

We have met with large success in the Hawaiian Islands and Porto Rico during the past five years as is evidenced by the fact that the following plantations, carrying the largest tonnage of cane at the LEAST COST PER TON, are using our portable track and cane-cars for the service. We refer with pleasure to the list below of a few users of our equipment:

PARTIAL LIST OF USERS OF THE "GREGG" EQUIPMENT. LISTA DE ALGUNOS INGENIOS EN LOS QUE SE USA EL EQUIPO "GREGG."			Miles of "Gregg" Portable Track in use. "Millas de Vía portátil Gregg" que se hallan en uso.	Number of "Gregg" Cane-cars in use. Número de Vagonetas "Gregg" que se hallan en uso.
Ewa Plantation Co.,	Ewa,	Hawaii,	5	
Waialua Agricultural Co.,	Waialua,	"	12	550
Kahuku Plantation Co.,	Kahuku	"	3½	100
Waianae Co.,	Waianea,	"		25
Oahu Sugar Co.,	Waipahu,	"	4	
Honolulu Plantation Co.,	Aiea,	"	7	100
Puna Sugar Co.,	Kapoho,	"	3½	
Olaa Sugar Co.,	Olaa,	"	3½	
Hawaiian Commercial & Sugar Co.,	Puunene,	"	12	575
Paia Plantation,	Paia,	"	3	130
Haiku Sugar Co.,	Hamakuapoko,	"	3	100
Kihei Plantation Co.,	Kihei,	"	6	235
Gay & Robinson,	Makaweli,	"	1	
Hawaiian Sugar Co.,	Makaweli,	"	3½	
Makee Sugar Co.,	Kealia,	"	1	
Waimea Sugar Mill Co.,	Waimea,	"	1	
Central Aguirre,	Aguirre, Porto Rico,		3	300
Lehue Plantation,	Kauai, Hawaii,		1	

"Gregg Steel" Wheels as Used on All "Gregg" Cars.

Ruedas de "Acaro Gregg" usadas en todas la vagonetas "Gregg."

"GREGG PATENT" PORTABLE TRACK.

Patented Dec. 12, 1899.

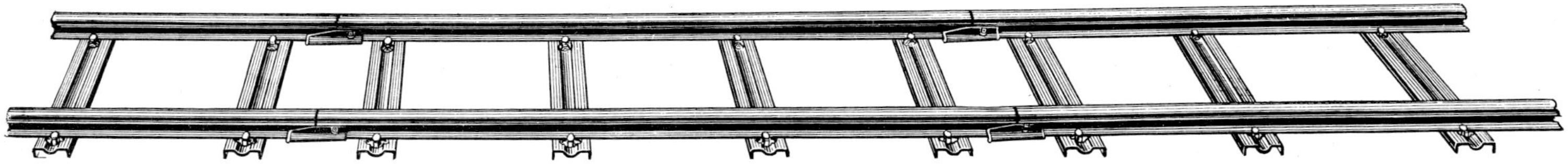

TOP VIEW OF TIE.
TRAVIESA, PARTE SUPERIOR.

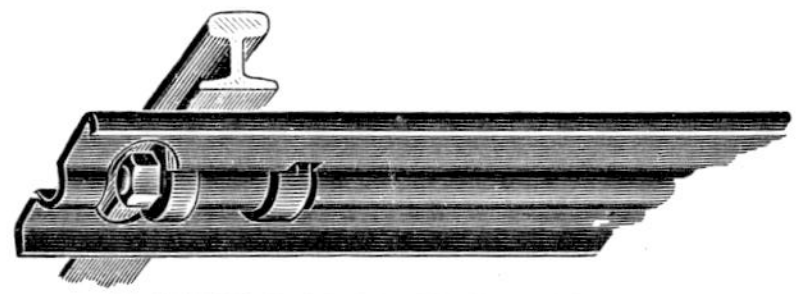

BOTTOM VIEW OF TIE.
TRAVIESA, PARTE INFERIOR

CLIP AND BOLT.
ABRAZADERA Y PERNO.

"*Gregg Patent*" Ties are made of special analysis steel, rolled on special rolls, the shape combining maximum stiffness and bearing surface with minimum weight Clips are made of best quality malleable iron. Bolts of steel with special side-head. The clips grip both edges of the rail and tie firmly, holding them perfectly solid with absolutely no chance of working loose. *No Rivets* to break off and consequent spreading of rails. Our track is so compact that it can be transported much more cheaply than track made with wooden ties and its life is more than five times as long.

We furnish "Gregg" ties any width of gauge and for any weight of rail.

"GREGG PATENT" SPLICE SHOES.

Patented Dec. 18, 1900.

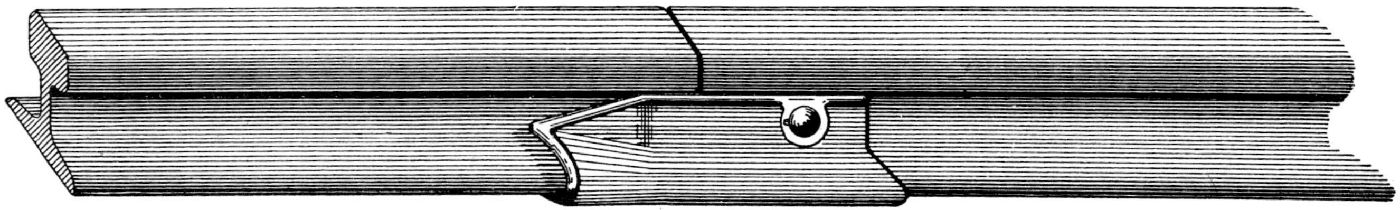

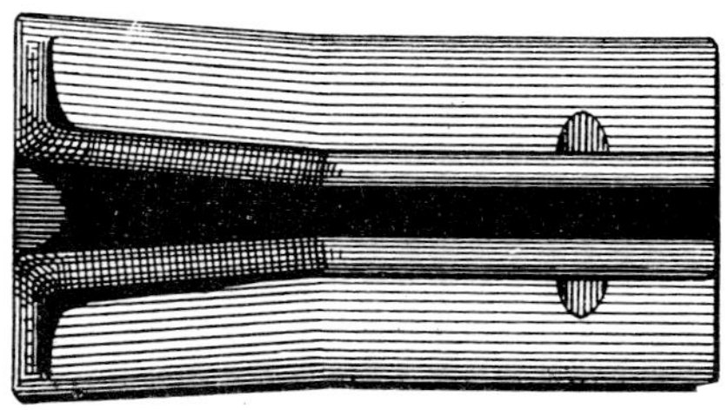

Made of the best quality of MALLEABLE IRON. Splice Shoes are bolted or riveted PERMANENTLY to ONE end of each section of portable track. The free rail-ends at other end of section slide into the mouth of shoes easily, and are supported rigidly and held in perfect line with the rail, forming a perfect rail joint.

When track is put together or taken apart no bolting or unbolting is necessary. NO WRENCH IS REQUIRED.

This track can be laid and taken up as fast as it can be handled.

We furnish shoes to fit ANY size rail.

When once assembled, each section of "GREGG" portable track becomes practically one solid piece. There are no loose parts such as clutch bolts, clamps, etc., to be lost. By a section we mean two rails, three, four or five ties, (according to length of rails) two splice shoes and the necessary clips and bolts.

Showing method of transporting "Gregg" portable track from one field to another, at the plantation of the Hawaiian Commercial and Sugar Company, Puunene, Maui, Hawaii.

El dibujo muestra la manera de transportar Via Portátil "Gregg" de una siembra de caña á otra, en la Hacienda de la Hawaiian Commercial Sugar Company, Puunene, Maui, Hawaii.

Piles of "Gregg Patent" portable track on estate of the Honolulu Plantation Company, Aiea, Hawaii.

Montones de Vias Portátiles "Patente Gregg" en la Hacienda de la Honolulu Plantation Company, Aiea, Hawaii.

TIE SECTION.

Straight section of portable track fitted with splice shoes.

Sección derecha de via portátil con uniones ajustadas.

Weight of Sections.

Weight per section of track, 36″ gauge, 5 meters long, complete with ties and joints.

Peso de los Tramos.

Peso por tramo de vía de 36 pulgs. de ancho y 5 metros de largo, completo, con durimentes y uniones.

Weight of Rail per Yard. El Peso del Rail por Yarda.	Weight of Sections. Peso de los Tramos.
8 lb. rail	145 lbs.
10 lb. rail	167 lbs.
12 lb. rail	215 lbs.
14 lb. rail	233 lbs.
16 lb. rail	259 lbs.
20 lb. rail	304 lbs.

Specifications for Hawaiian Type "Gregg" Sugar Cane Car.

FRAME: Length, 11 feet; width, 6 feet (inside dimensions.)

Five sills, 4 x 8 inches; 2 end cross sills, 6 x 8 inches; all mortised and tenoned and securely held together by two ¾ inch, and two 1⅛ inch steel rods running through entire length of Car, and three steel cross rods.

Stakes supporting upright ends of Car are securely bolted to sills and extend down the entire depth of sills instead of being set on top, thus forming one of our lightest but strongest features.

WHEELS: "Gregg Steel;" diameter, 18 inches; tread, 3¼ inches.

BRAKE: Lever Pattern, very powerful, acting on all four wheels, and can be operated from either side of Car.

JOURNAL BOXES AND PEDESTALS: "Gregg Pattern," Malleable iron, with removable solid brass journal-bearings, and 8 steel coil Car springs.

STAKE POCKETS: "Gregg Patent," Malleable iron, holding 3x4 inch stake.

DRAWHEADS: "Gregg Patent," Malleable iron, held in place by 1⅛ steel rods, running through entire length of Car frame.

These drawheads are practically one piece, coupling links and pins being attached PERMANENTLY. A heavy spring, made of ⅞ inch steel, eases both buffing and pulling.

NUTS: All frame rods, brake-rods and bolts holding running gear to frame, have patent Elastic Self-Locking Steel Nuts, whieh never work loose, but may be readily tightened at any time.

Above dimensions will be altered to suit Cars of any gauge or of any capacity desired.

SPECIFICATIONS FOR HAWAIIAN TYPE "GREGG" SUGAR-CANE CAR WITH HINGED FLARING SIDES.

DESIGN No. 2.

This car takes same specifications as our side-stake car described on page 16 except that flaring side doors are used instead of stake pockets. Each door is supplied with three heavy strap hinges made of 2x⅜ steel, bolted on.

The doors are held closed by means of ring chain locks at upper corners, which hold them securly and are easily released when desired.

NO. 1 98.

Code word "STATUTABLE" Palabra de la clave.

STEEL FRAME 5 TON CANE CAR.

With hinged flaring side doors.

Standard size of frame 6 x 12 feet.

Any gauge.

Full "Gregg Standard" equipment.

Made with perpendicular side doors if desired.

CARRO DE 5 TONELADAS CON ARMAZÓN DE ACERO.

Carro con compuertas abisagradas. El tamaño del armazón es de 6 por 12 piés.

Se fabrica para via de cualquier ancho.

Lleva equipo completo "Modelo Gregg." Se fabrica con compuertas perpendiculares si se desea.

Same as No. 198 above, showing door open.

Lo mismo que el número 198 mostrado en ésta página, pero con la compuerta abierta.

No. 205.

Code word "STATUTO" Palabra de la clave.

STANDARD 5 TON STEEL FRAME CANE CAR.

Full "Gregg Standard" specifications.

Standard size of frame, 6 x 12 feet.

Particular attention has been given to the design and construction of this car, and we do not hesitate to state that it is the best 4-wheel 5-ton cane car on the market.

We have thousands of them in use in all cane sugar countries.

Steel wheels, malleable iron *Journal Boxes and Pedestals*, "Gregg" Patent malleable iron spring couplers and quick releasing stake pockets.

CARRO MODELO DE 5 TONELADAS PARA CAÑA, CON ARMAZÓN DE ACERO.

Especificaciones completas "Modelo Gregg."

Tamaño modelo del armazón 6 por 12 piés.

Mucho cuidado se ha puesto en el trazado y construcción de éste carro, y no vacilamos en decir que es el mejor carro de 4 ruedas y de 5 toneladas en el mercado.

Tenemos miles de ellos en uso actual en todos los paises azucareros.

Ruedas de acero, chumaceras y pedestales de hierro maleable, Topes de hierro maleable "Patente Gregg" y cajas para estacas de pronto desprendimiento de hierro maleable también.

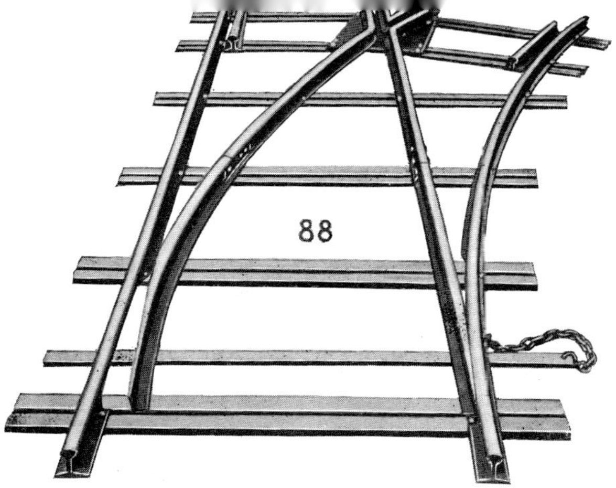

No. 88
Right Hand Switch.
Without lever throw.

Desvio á la Derecha.
Sin palanca de cambio.

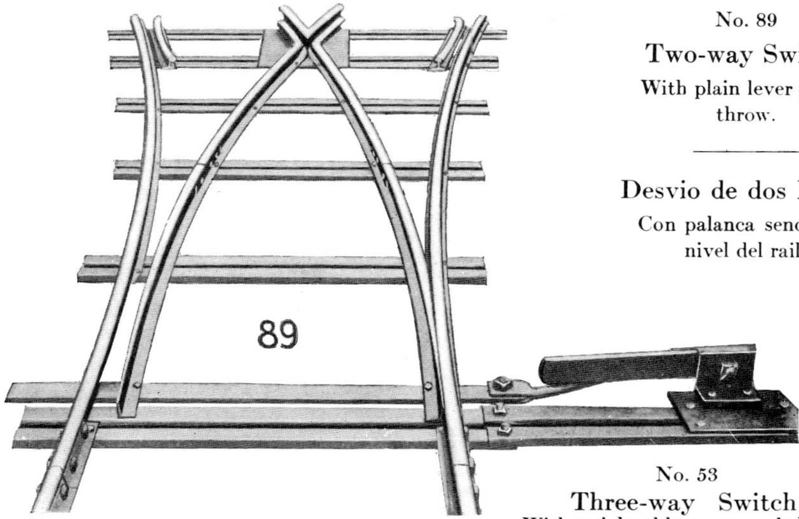

No. 89
Two-way Switch.
With plain lever ground throw.

Desvio de dos Lados.
Con palanca sencilla al nivel del rail.

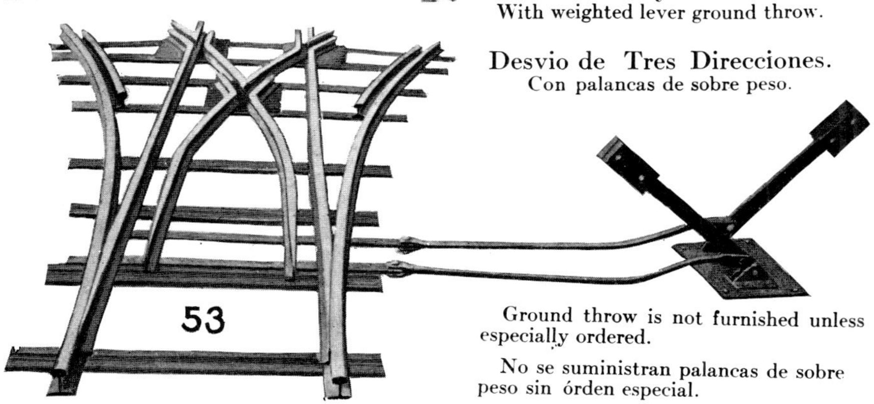

No. 53
Three-way Switch.
With weighted lever ground throw.

Desvio de Tres Direcciones.
Con palancas de sobre peso.

Ground throw is not furnished unless especially ordered.

No se suministran palancas de sobre peso sin órden especial.

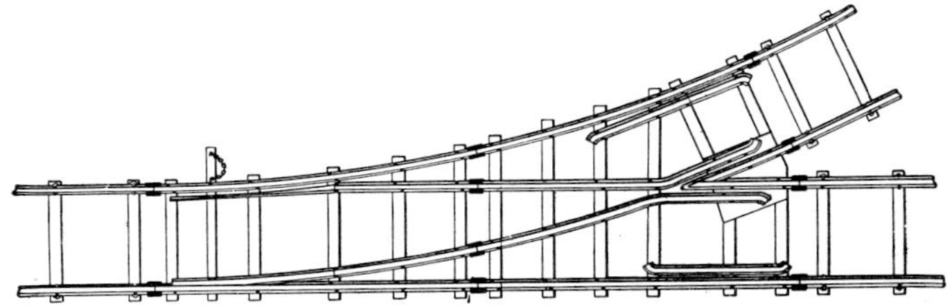

No. 234

Two-piece Portable Switch, Left Hand.

These switches are joined in the middle with four "Gregg" Patent Splice Shoes.

Weight, 300 to 1,000 pounds, according to length and weight of rail.

Two advantages in two-piece splice shoe switch: 1st, A longer, easier curve; 2nd, It is handled in two parts with less labor than a one-piece switch.

We recommend two-piece switches for plantation or any out-door use.

Desvio Portátil en Dos Piezas Con Cambio á La Izquierda.

Estos desvios están unidos en el medio por cuatro uniones "Gregg."

Peso de 300 á 1,000 libras, según la longitud y peso de los railes.

El desvio de dos Piezas con uniones tiene dos ventajas: 1a, Una curva más larga y más fácil; 2a, Se opera con mayor facilidad siendo de dos piezas que de una.

Recomendamos el desvio de dos piezas, para plantaciones ó cualquier uso exterior.

Climbing Switch.

MAIN LINE OPEN.

LINEA PRINCIPAL ABIERTA.

Allows passage of trains on main line without removing entire switch, but throwing points only.

Desvio de Ascencion.

MAIN LINE CLOSED.

LINEA PRINCIPAL CERRADA.

Permite el paso de un tren sobre la vía principal sin tener que mover el cambiavía; sólo teniendo que echar los puntos.

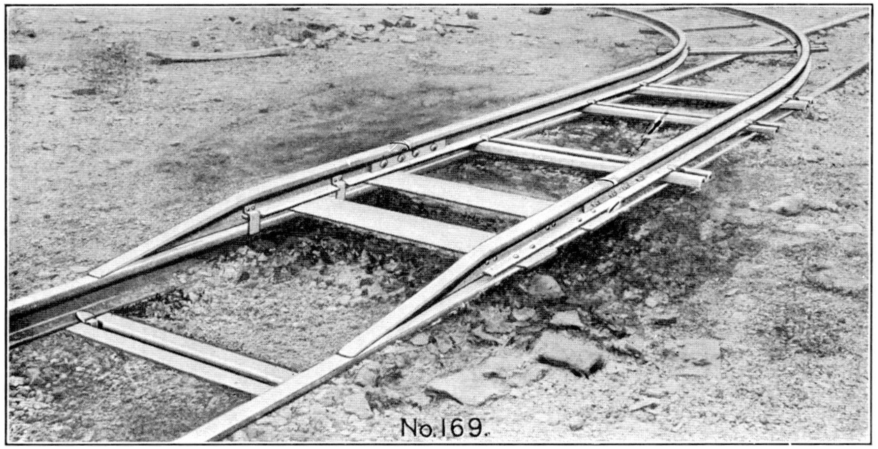

Code word "*SYNEPY*" Palabra de la clave.

Track Incline.

Made any gauge and length (usually 4½ feet), and of any weight rail. Incline includes that portion of pointed rails in illustration, to center of fish plates, which hold it to adjoining curved section of track.

Cambiavia Sobrepuesta Con Declive.

Se construyen de cualquier ancho y largo (tamaño corriente 4½ pies de largo), para railes de cualquier peso.

El declive representa la parte de los railes inclinados hasta las uniones como se demuestra en nuestro dibujo.

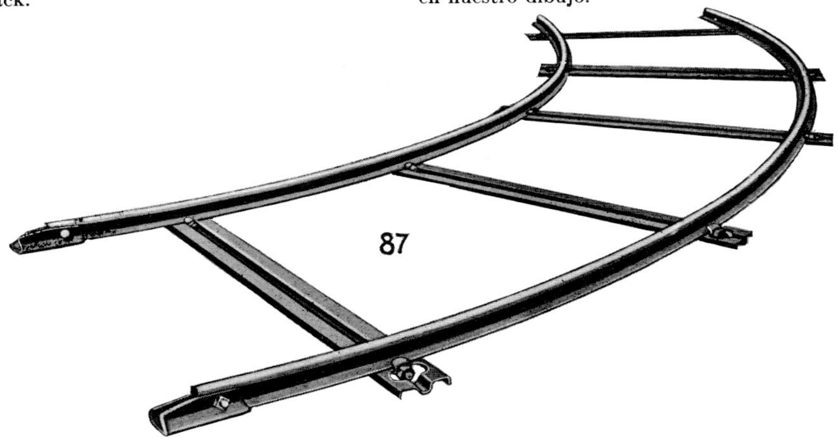

No. 87

Portable Curved Section.

Made any radius, gauge and length.

Furnished with splice shoes as shown above, or fish plate joints.

Sección Curva de Via Portátil.

Se fabrican de cualquier radio, ancho y largo.

Las suministramos con uniones como se ven en el grabado de arriba, ó con empalmes.

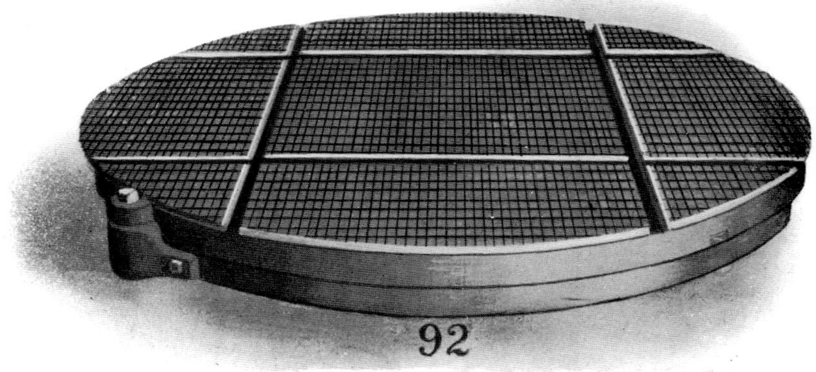

Code word "*SYNERGISTS*" Palabra de la clave.

Cast Iron Ball Bearing Turntable.

Equipped with automatic lock. Diameters, 4, 5 and 6 feet. Tracks any gauge.

Plataforma Giratoria.

Plataforma de Hierro Fundido, con bolines. Equipada con cerradura automática. Diametro, 4, 5 y 6 pies. Vias de cualquier ancho.

No. 20

Code word "*SYNERGUM*" Palabra de la clave.

Turntable.

Steel Turntable with Rail Top. Made any diameter and gauge.

Plataforma Giratoria de Acero.

Plataforma giratoria con railes en la parte superior. Construida de cualquier diametro y ancho.

Code word "*STAUBHAAR*" Palabra de la clave.

Single Track Transfer Car.

With roller bearings. Made any gauge and capacity. Furnished with two tracks if required.

Carro de Cambio, de Via Simple.

Se fabrica con bolines. De cualquier ancho y capacidad.

' GRECG'' STANDARD DIAMOND ARCH BAR TRUCK.

All steel and malleable iron construction, no cast iron parts.

Made any gauge and capacity, with wheels 18" to 33" in diameter.

CARRETILLA MODELO "GRECG" CON LA BARRA DE UNIÓN EN FORMA DE DIAMANTE.

La construccion de esta carretilla es enteramente de acero y hierro maleable; no hay ninguna pieza de hierro fundido.
Se construye de cualquier capacidad y para via de cualquier ancho. Con ruedas de 18 á 33 pulgadas de diametro.

DIAMOND FRAME TRUCK WITH WOOD BOLSTERS AND SPRINC PLANK.

All metal parts of steel and malleable iron.
Made any capacity and gauge, with wheels 18" to 33" in diameter.

CARRETILLA CON ARMAZÓN EN FORMA DE DIAMANTE CON TRAVESAÑO DE MADERA Y TABLON CON MUELLES.

Todas las piezas de metal de ésta carretilla son de acero y hierro maleable.
Se construye de cualquier capacidad y para via de cualquier ancho. Con ruedas de 18 á 33 pulgadas de diametro.

No. 187.
Code word "SYNODAL" Palabra de la clave.

"GRECG PATTERN" ALL METAL TRUCK.

Malleable iron journal boxes and pedestals, the latter enclosing two coil springs for each box.
Made any gauge.
Built for 8 to 15 ton capacity cars, with wheels 16" to 24" in diameter.

CARRETILLA "MODELO GRECG" TODA DE METAL.

Las chumaceras y los pedestales son de Hierro Maleable; los pedestales llevan dos muelles para cada chumacera.
Se fabrica para via de cualquier ancho. Ésta carretilla se construye para 8 ó 15 tonelas de capacidad con ruedas de 16 á 24 pulgadas de diámetro.

No. 259

MALLEABLE IRON JOURNAL BOXES.

The above illustration shows type of box used o trucks Nos. 23 and 63 for cars of from 15 to 30 tons capacity.
We furnish Malleable Iron Journal Boxes for any make of cars, to interchange with and replace Cast Iron Boxes.

CHUMACERAS DE HIERRO MALEABLE.

El grabado de arriba muestra el estilo de chumacera usada en las carretillas Nos 23 y 63 para carros, desde 15 toneladas á 30 toneladas de capacidad.
Suministramos Chumaceras de Hierro Maleable para carros de cualquier marca, y reemplazamos chumaceras de Hierro Fundido.

"Gregg" Double Truck Flat-Car.

DESIGN No. 5.

FRAME: Length, 24 feet; width, 6 feet (inside dimensions). Four lengthwise sills, 4x7½ inches, mortised and tenoned. Two end crosswise sills, 6x7½ inches, mortised and tenoned. Two truss sills, 3x6 inches. Bolsters 4x12 inches. Four lengthwise truss rods, ¾ inch, with wrought iron turn buckles.

TRUCKS: "Gregg" pattern, with wheels, journal boxes, pedestals, springs and brake, as described on page 16. All bolster plates malleable iron.

DRAWHEADS: "Gregg" patent, malleable iron. Nuts on all rods and running gear are patent "Elastic Self-Locking."

NOTE: No cast-iron whatever on any of these cars save brake-shoes.

Above dimensions will be altered to suit cars of ANY gauge, or of any desired capacity, up to 20,000 lbs.

Drawheads may be attached at ANY desired height.

Ratchet wheel-brake furnished for one or both ends, instead of lever brake, if desired.

"Gregg" patent malleable iron stake-pockets to hold side stakes, of 3x4 inch lumber, furnished if desired. Any number of stakes may be used. Furnished with or without upright ends; also with removable or hinged solid board sides.

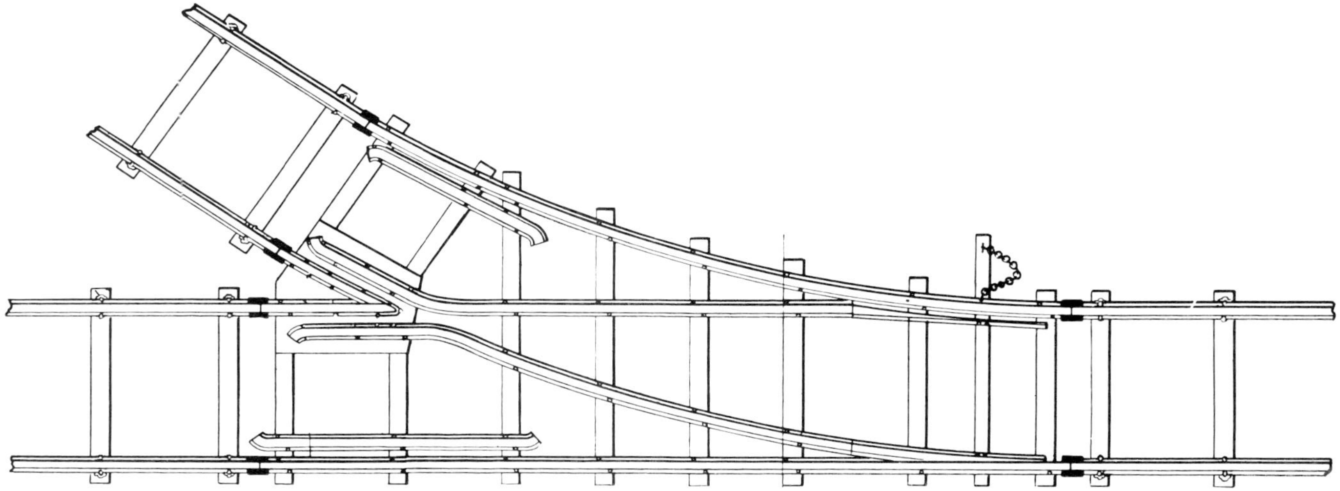

ONE PIECE PORTABLE SWITCH, RIGHT HAND.

UNA PIEZA CAMBIAVÍA PORTÁTIL—CAMBIO Á LA DERECHA.

Weighs five hundred to seven hundred pounds, according to length and weight of rail. We furnish above switches RIGHT OR LEFT HAND, in any length or gauge, or of any weight rail desired. Cross ties of same steel tie-section as shown on page 5, securely riveted to rails. Frogs and points forged from steel rail.

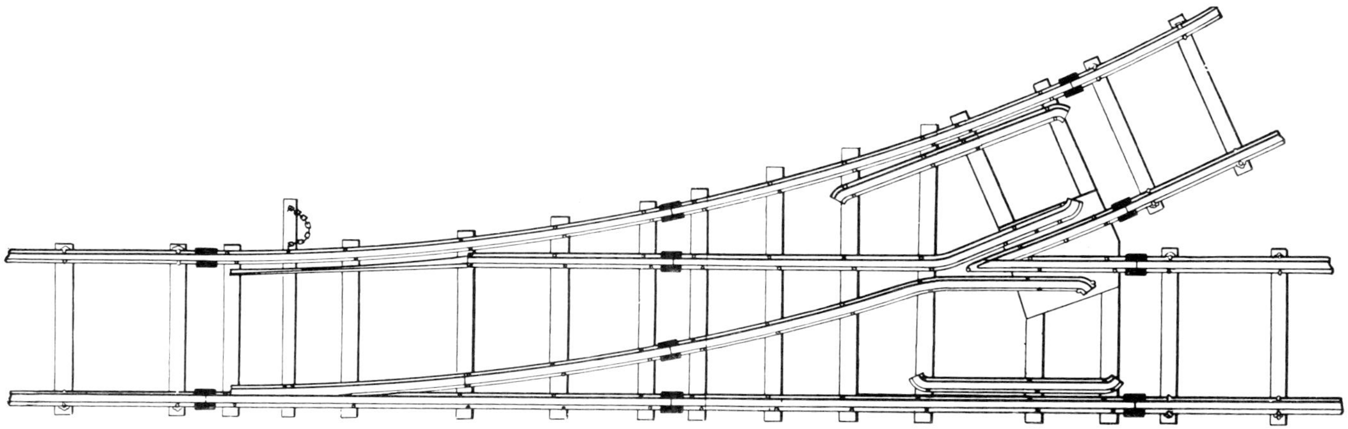

TWO-PIECE PORTABLE SWITCH, LEFT HAND.

CAMBRAVÍA PORTÁTIL EN DOS PIEZAS—CON CAMBIO Á LA IZQUIERDA.

These switches are joined in the middle with four "Gregg" Patent Splice Shoes.

Weighs six hundred to one thousand pounds, according to length and weight of rail.

TWO ADVANTAGES IN SPLICE SHOE SWITCH: 1st. A longer, easier curve. 2nd. It is handled in two parts with less labor than a one piece switch.

We furnish above switches RIGHT OR LEFT HAND, in any length or gauge, or any weight rail desired. Cross ties of same steel tie-section as shown on page 5, securely riveted to rails. Frogs and points forged from steel rail.

SPECIFICATIONS FOR "GREGG" FOUR-WHEEL FLAT CAR FOR HAULING PORTABLE TRACK, RAILS, LUMBER, ETC.

DESIGN No. 3.

This car takes same specifications as our Hawaiian Type Car, described on page 16, omitting the upright end stakes, and is equipped with a horizontal motion, powerful brake on all four wheels, worked from the side as shown.

Dimensions will be altered to suit any gauge or capacity.

Car can be furnished with or without drawheads and brake, if so desired.

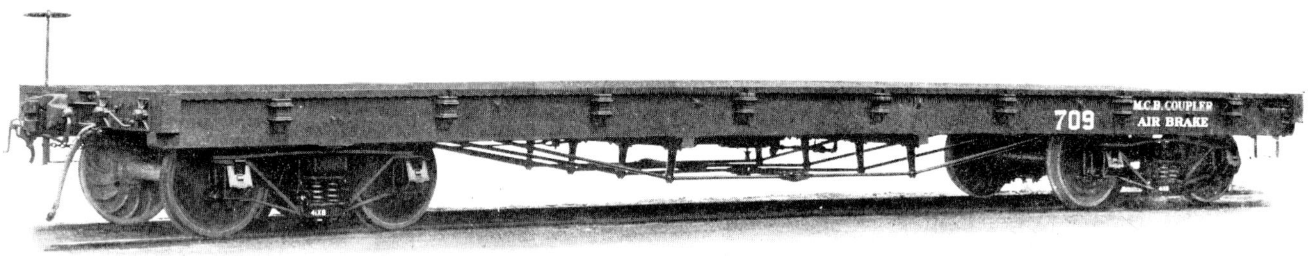

SPECIFICATIONS FOR "GREGG" DIAMOND-FRAME DOUBLE TRUCK CAR.

DESIGN NO. 6.

FRAME: Length 33 feet; width 6 feet. Lengthwise sills 4 x 8 inches; End cross sills 6 x 10 inches, mortised and tenoned. Truss rods $\frac{7}{8}$ to $1\frac{1}{4}$ inch, depending upon capacity, with wrought-iron turn buckles.

TRUCKS: Standard Diamond Frame, wood bolsters or all steel. Wheels 20 to 30 inches diameter. Axles $2\frac{1}{2}$ to $3\frac{1}{2}$ inch diameter.

DRAWHEADS: Automatic M. C. B. couplers, or common link and pin type, provided with heavy springs.

NUTS: On all rods and running gear, patent "Elastic Self-Locking" steel

---

We will furnish this design of car in any desired dimensions. Ratchet-wheel brake furnished for one or both trucks, operated from one end of car. Any desired number of side or end stake-pockets, solid board upright ends, or hinged or solid sides will be furnished.

GREGG CAR PARTS, ALL MALLEABLE IRON.

JOURNAL BOX: Our Spring Lid Journal Box has proven a great trade winner on account of the device used for holding Brass in place. All users of light railway equipment on portable track know that cars are frequently off the track on account of uneven ground and careless handling, and with the ordinary journal box and brass, great difficulty is experienced on account of brasses turning and falling into bottom of box when the cars are derailed and unless great care is taken to properly replace brass, the journal box is quickly worn through on the top.

THE "GREGG" BOX is so constructed that it is absolutely impossible for the brass to get out of its proper position This feature gives the users of our cars the advantage of removable brasses without the disadvantages which the ordinary box offers.

The design of our box gives DEEP OIL AND WASTE STORAGE and the condition of journal is always in plain view.

PEDESTAL: Our Malleable Pedestal is designed to carry ten times the weight it is called upon to carry in Cane Car service, yet it is very light and neat in appearance. It is so designed that it entirely surroudds the Springs, thus protecting them, as well as forming a very solid construction.

We supply the "Gregg" Pedestal and Journal Box to fit any car and make them to interchange with old parts now in use.

STAKE POCKETS: Our "Gregg" Patent Stake Pocket was designed to take the place of the old solid pocket from which the stake can be removed only by great effort and loss of time. The door of the "Gregg" pocket is released by a slight tap on the latch allowing stake to be readily removed. Pocket is made to carry 3x4 inch Rough stake without any trimming, and a shelf in the bottom prevents stake from falling through. This pocket is made with or without Hook for animal towing as shown in illustration.

"Gregg" Patent Malleable-iron Journal Box and Pedestal resting on Pedestal strap.

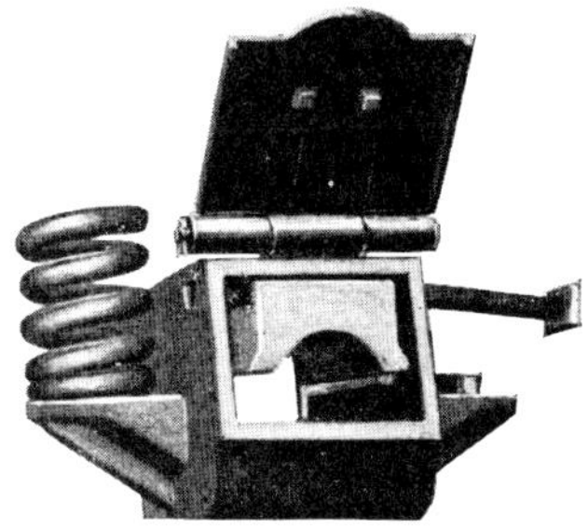

Showing box with pedestal removed, lid opened and brass nearly withdrawn. One Spring is shown on its Bracket at left side of box.

"Gregg" Patent Stake Pocket.

Pocket in center is shown with door or lid open.
Pocket on the left shows Left Hand Animal Towing Hook.
Pocket on the right shows Right Hand Animal Towing Hook.

"GREGG" DRAWHEAD.

FOR LIGHT AND INDUSTRIAL RAILWAYS.

Patented Sept. 11, 1900.

On the sugar plantations of the Hawaiian Islands, where there are operated on each estate from five to thirty miles of permanent railways, with five to fifteen miles of portable-railway in conjunction, and from one hundred to seven hundred sugar-cane cars, with average carrying capacity of six tons, there arose a demand for a coupler that would prevent the loss of coupling links.

This demand was vital, not alone because of the delay and expense of links being missing when a train of cane-cars was to be coupled up, but at the mill great difficulty was experienced in the use of ordinary bumpers, with loose links and pins, on account of the links and pins falling into the cane conveyor and being carried through the mill, causing great damage to crushers and rolls.

Automatic M. C. B. couplers, such as are used on all standard railways, were tried and found useless, as they would not remain interlocked or coupled on the rough portable tracks and were a considerable hindrance on the sharp curves.

We have met and successfully overcome the situation by perfecting our "Gregg" coupler described below.

The coupler is extremely simple, consisting principally of two malleable-iron castings, a socket held entirely outside of end sill of car, (by the lateral tie or truss rods passing entirely through car-frame to the socket on the other end of the car) in which is inserted the head or bumper, in a manner allowing it to play back and forth, pulling or pushing, as the case may be, against a heavy steel coil spring, also contained in the socket. The two castings are held together by a square steel pin passing vertically through both.

One coupling link is held permanently in the oblique slot in each bumper-head, by an oval head-pin which is enlarged at the bottom so that it cannot be withdrawn from the slot but works freely laterally in it.

Thus our complete drawhead, which consists of only seven pieces, when assembled becomes practically *one* piece, no part of which can be lost. It is all in plain sight and may be attached or removed from a car by simply applying or removing the two nuts on the truss rods, which hold socket against end sill of car.

We have thousands in use, all giving universal satisfaction.

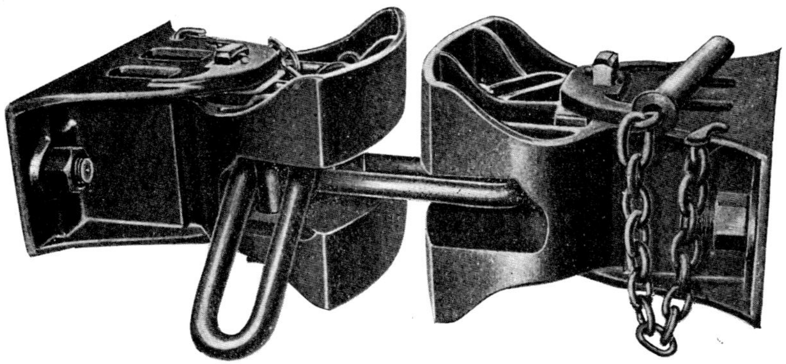

"GREGG" PATENT CANE UNLOADERS.

As installed at mill of The Wailuku Sugar Co., Hawaii.

DESCARGADORES DE CAÑA PATENTE "GREGG."

Según instalados en la hacienda Wailuku Sugar Co., Hawaii.

"GREGG" PATENT CANE UNLOADERS.

As installed at mill of The Pioneer Mill Co., Hawaii.

DESCARGADORES DE CAÑA PATENTE "GREGG."

Según instalados en la hacienda Pioneer Mill Co., Hawaii.

STEEL WHEEL VELOCIPEDE.

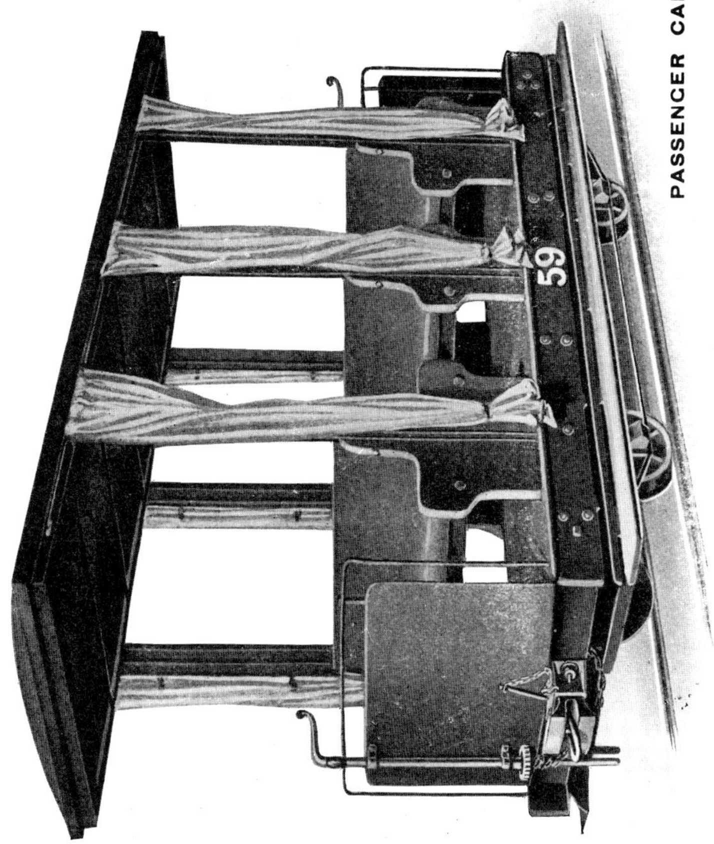

PASSENGER CAR.

CANOPY TOP VELOCIPEDE.

"GREGG" CANE-CAR UNLOADER.

(Froelich Type)

In use at Hanamaulu Mill and Lihue Plantation Co., Lihue, Hawaii, and Makee Sugar Co., Kealia, Hawaii.

Showing a machine with two lines of chain connected by cross bars carrying rakes or teeth, unloading cars equipped with hinged flaring sides or doors of solid boarding.

TWO "GREGG" CANE-CAR UNLOADERS.

At Puunene Mill of Hawaiian Commercial & Sugar Co., Puunene, Hawaii.

Showing machines with four triangles and four lines of rake chain to each car.

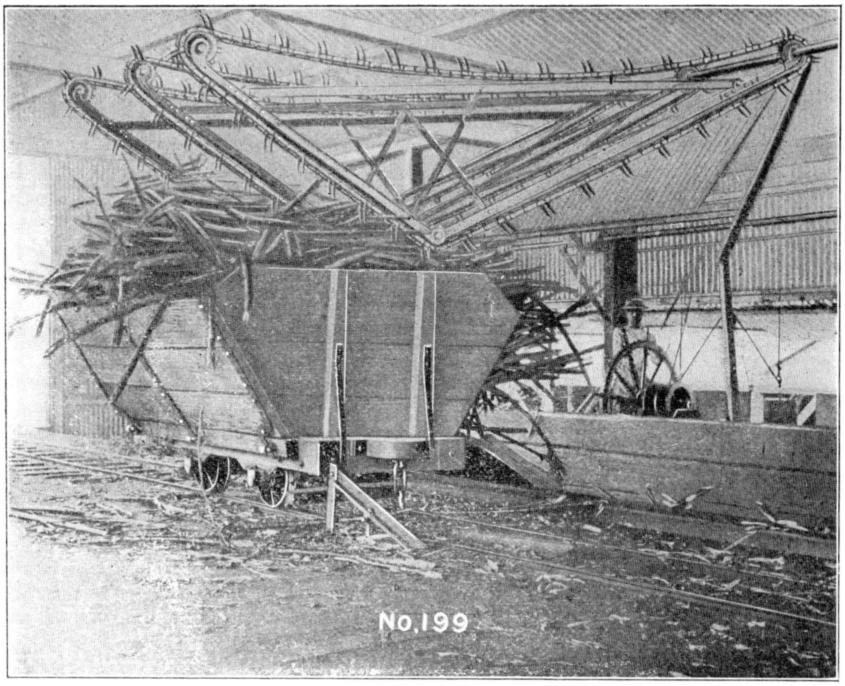

"GREGG" PATENT CANE UNLOADER.

As installed at mill of Kilawea Sugar Plantation Co., Hawaii.

TWO "GREGG" CANE-CAR UNLOADERS.

At Waialua Agricultural Co., Waialua, Hawaii.

Showing both machines elevated, leaving no obstruction whatever to tracks or cane carrier.

Testimonials.

Ewa, Hawaii, Feb. 27th, 1904.

The Gregg Company, Ltd.
Newburgh, N. Y.

Dear Sirs:—Your cane unloaders have been in use at the mills of the Ewa Plantation Co. for several years past, and have proved to be very efficient machines. They have reduced the expense of unloading cane at the cane-carrier very materially, and are much more satisfactory in every way than unloading by hand, particularly because the mills can always be kept properly fed with cane. This is an item fully as important as the saving of labor.

I take pleasure in testifying to their economy and efficiency. The feed supply can be regulated to suit the needs of the mill. Two machines placed one on each side of carrier can deliver any amount required to 1,250 tons in 24 hours.

Yours truly,
(Signed) GEO. F. RENTON,
Manager Ewa Plantation Co.

Manager Geo. R. Ewart, of the Kilauea Sugar Plantation Co., of Hawaii, wrote us under date of January 24, 1902: "I am pleased to say that your cane-car unloading machine is doing its work splendidly. I have never had anything please me so much; besides it saves us twelve men per 24 hours."

The Paia Plantation of Hawaii writes as follows, dated February 12, 1901: "We are pleased to advise you that we have had your cane unloading machine in operation for the last three years, and we find that it has effected a saving in this department of about 40 per cent. in labor."

(Signed) D. C. LINDSAY, Manager.

Mr. Goodale, manager of the Waialua Agricultural Co., of Hawaii, writes as follows: "I would say that we put in two of your cane unloaders, four triangles to each car, and they have now been in operation two weeks. The cane carrier work is now being done by eight men on each shift, including the pulling out of car stakes, pushing in loaded cars, pushing out empty cars, running the cane unloaders and keeping the cane yard clean. Before putting in the cane unloaders it required thirteen men on each shift to do this work."

Kealia, Hawaii, February 26th, 1903.

The Gregg Company, Ltd.
Honolulu.

Dear Sirs:—I am enclosing herewith our check for $2,751.50 in settlement of our account. You will remember that we were due to pay for the cane-car unloader on the 1st of March, provided it proved satisfactory. We are saving over $9.00 a day by using the unloader, and in addition considerable worry and annoyance from the men. We did very cheap work by hand, and I hardly thought we could do as well with the machine as we have. I believe your "Froehlich Type" machine the best for our flaring side cars.

Yours truly,
MAKEE SUGAR CO.,
(Signed) Geo. H. Fairchild, Manager.

Guatemala, March 30th, 1904.

The Gregg Company, Ltd.
Newburgh, N. Y.

Dear Sirs:—It gives us pleasure to inform you that your "Gregg" Cane-car Unloader, which we installed ourselves, following the blue-prints which you sent us, is giving the most satisfactory results. It works well and does not give us the slightest trouble.

It will always be very agreeable for us to correspond with you whenever we may be in need of any materials for our railway or other machinery.

Yours very truly,
(Signed) HERRERA & CO.

Chapter Seven / Glimpses of the Common Carriers

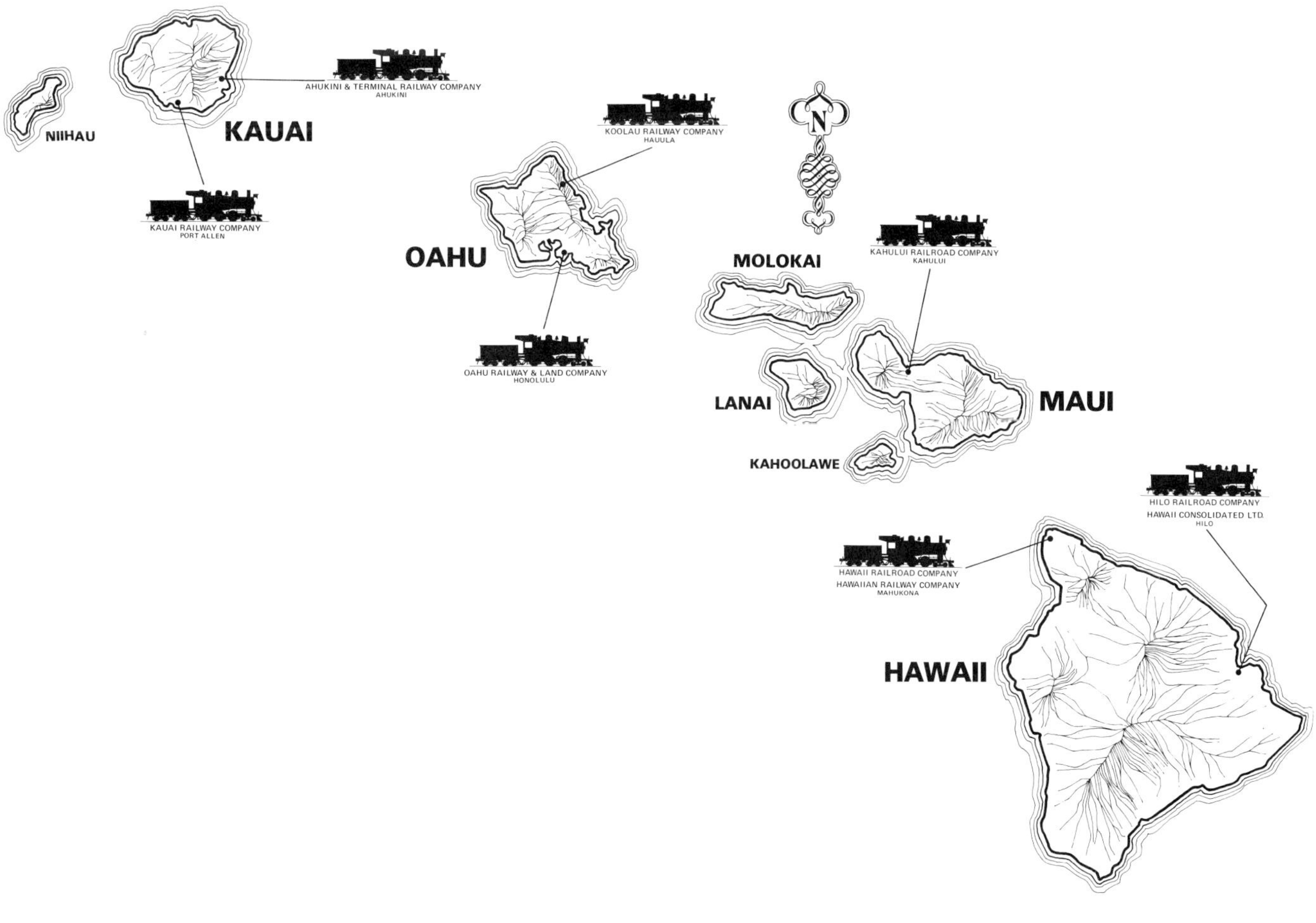

AHUKINI TERMINAL & RAILWAY COMPANY
—LIHUE, KAUAI—

Aerial view of A.T.& Rwy Co. Ahukini Landing, Hanamaulu Harbor. Circa 1925. *Amfac Inc.*

Early view of A.T.& Rwy. Co. track passing under the Hanamaulu Bridge about 1922. *Amfac Inc.*

Later day view of Hanamaulu River bridge. *Amfac Inc.*

H. K. Porter locomotive "Kalalau," Ahukini Terminal road locomotive #1 on the rip track at Lihue Plantation 1947. *Fred Stindt photo, G.M. Best Collection.*

HAWAIIAN RAILROAD CO./HAWAII RAILWAY CO.
—MAHUKONA, HAWAII—

H.Rr.Co. "A Ke Ahi," Ransomes & Rapier switcher at Mahukona round house circa 1906. *Fred Denison Collection, H.S.P.A.*

Close-up view of "Kinau" at Mahukona in 1882. *John Boyle, Philco Ford.*

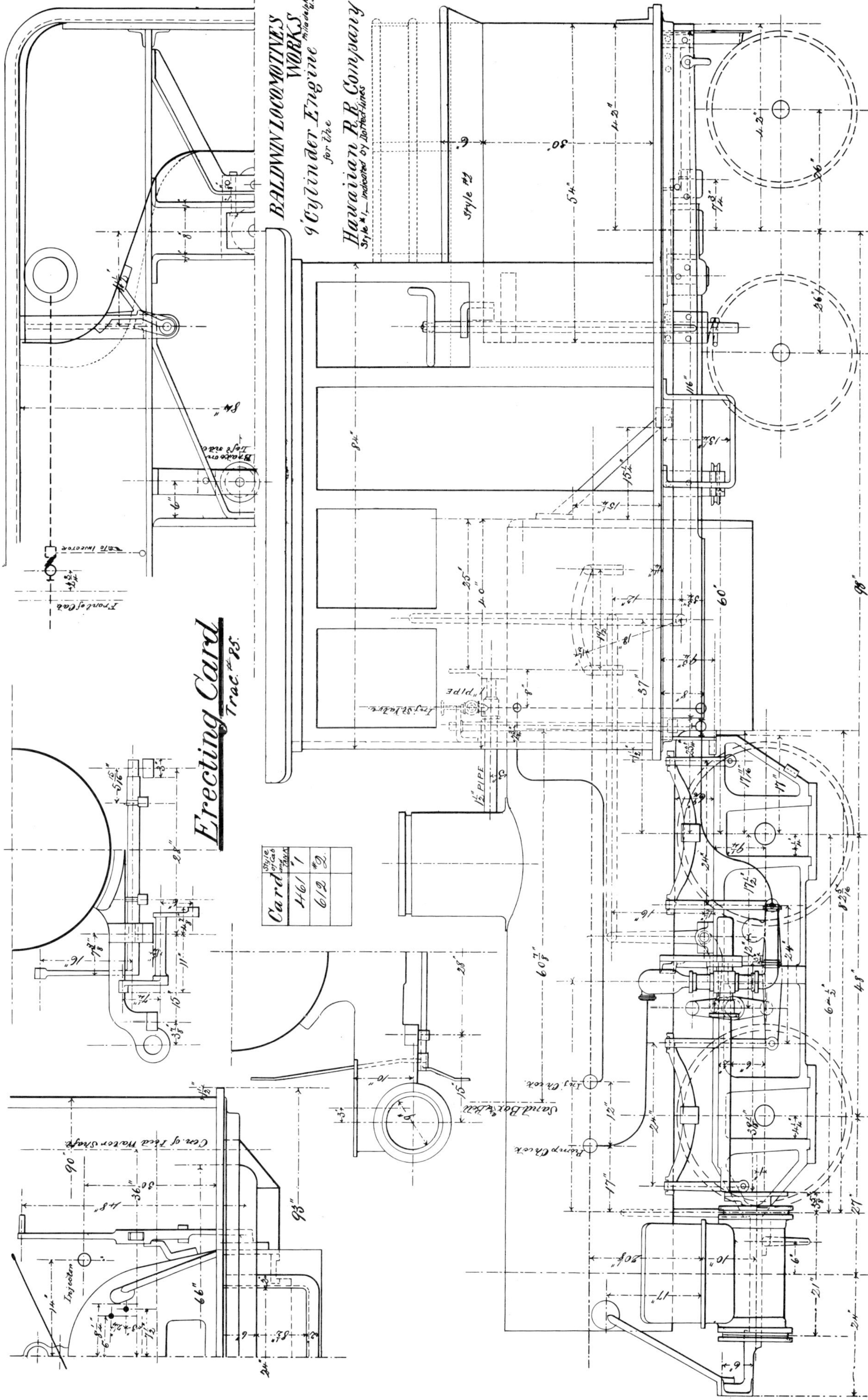

Locomotive "Kinau," later transferred to Kahului Railroad Co. Baldwin Class 8 11 1/3 C 2; Construction # 5640. Built 5/ 1881, 30-9x14. *Edward Bond Collection.*

"Kinau" passes "Beacon Point" near Mahukona with a train load of bag sugar in 1882. *A. A. Montano, Archives of Hawaii.*

Builder's photo of H.R.C. "Myrtle" #3, a Dickson Works Locomotive of 1900. *Alco Historic Photos.*

The "Leslie" H.R.Co. 2nd #1, once named "Kalakaua" works on the dock at Mahukona under the banner of Mahukona Terminals Ltd. 1941. *Baldwin Locomotives.*

Hawaii Railway Company #5 as restored to steam operation on the Tahoe, Trout Creek & Pacific Railroad, South Lake Tahoe 1971. *Iron Trails Photos — H. W. Brueckman.*

Mahukona terminal of the Hawaiian Railroad Company 1882. *A. A. Montano, Archives of Hawaii.*

Ruin of the Mahukona Boat House at end of track, Hawaii Rwy. Co. July 1969. *Author's Collection.*

The famed "Kalakaua Coach" at Mahukona in 1946. The finely etched glass panels in the door of this car are reported to be still in existence on the Island of Hawaii. *Baker Collection, Bishop Museum.*

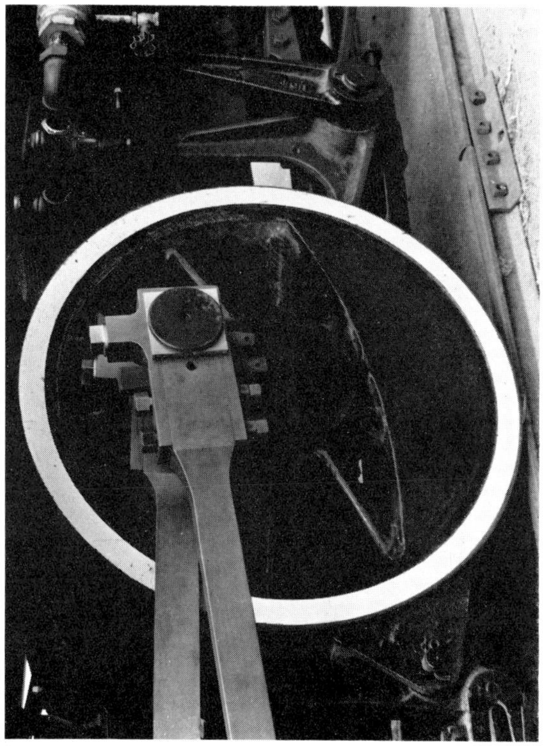

Classic view of Left Rear driver and rods of Hawaii Rwy. Co. #5, April 1970. *Dexter Pike.*

HILO RAILROAD CO. /
HAWAII CONSOLIDATED RAILWAY LTD.
—HILO, HAWAII—

Old Hilo Post Office and Railroad Depot at the foot of Waianuenue Avenue, Hilo Town. The palm tree at the left was protected by a county ordinance and was marked to show the height of water reached by the various Tsunamis (tidal Waves), that hit Hilo. View looking North. *Harry Wessell Collection.*

Front end view of H.Rr. #3 in roundhouse 1946. *Cochrane Collection, Bishop Museum.*

Front end view of H.C. Rwy., #5 in the roundhouse at Hilo. Originally an 0-6-0, this machine was converted to a 2-6-2 at the Waiakea shops. *Cochrane Collection, Bishop Museum.*

Hilo Railroad Co. #3 with train on the first section of the Hilo Breakwater about 1910. *Harry Wessel Collection.*

Ex Sierra Railway 0-6-0 as Hawaii Consolidated #5 in the rail yards at Hilo 1943. *G. H. Kneiss photo, G. M. Best Collection.*

Builder's photo of Hilo Railroad Co. #3, "Kilauea" 1902. *Baldwin Locomotive Works.*

Builder's photo of Hilo Railroad Co. #108, Baldwin built 1908. *Baldwin Locomotive Works, G. M. Best Collection.*

"Train from the North" — H.C. Rwy. Co., #15 crossing the Wailuku River bridge on March 5th, 1923, enroute to Hilo depot. *R. B. Jackson Collection.*

Gilbert H. Kneis photo of Hawaii Consolidated "ten wheeler", "Old Ninety Nine", ready for service in the Hilo yards, March 1943. *R. B. Jackson Collection.*

H.C.Rwy. Co. Baldwin locomotive, road engine #121 just arrived at Hilo 1921. *G. M. Best Collection.*

Road Engine #121 of the H.C.Rwy.Co. in rail service near the Waiakea roundhouse in 1946. *Cochrane Collection, Bishop Museum.*

Right hand view of H.C. Rwy. Co. #191 in 1946. *Vitaly V. Uzoff.*

Builder's photo of Alco-Cooke locomotive of the Hawaiian Consolidated Railway, road locomotive #196 1916. *G. M. Best Collection.*

Left hand view of H. C. Rwy. Co. #191 at Hilo in 1943. *G. H. Kneiss Collection.*

H. C. Rwy. Co. #192 near the water tank in the Waiakea yards 1943. *G. H. Kneiss, G. M. Best Collection.*

View of Hawaiian Consolidated "Galloping Goose" style Rail Bus, in the Waiakea yards in 1946. This style machine was a World War II passenger carrier. *Cochrane Collection, Bishop Museum.*

Hilo Railroad locomotive hauling rock from Puna quarry for the Hilo breakwater construction. Circa 1910. *Harry Wessel Collection.*

Hilo Railroad Co. #99 at Olaa, Hawaii about 1904. The mixed consist of an excursion open car, "Palace Car" coach, and box car demonstrates management's desires to accommodate all types of rail service. *Camera Hawaii.*

Old "99" leaving Hilo yards with a mixed consist in March 1943. *G. H. Kneiss photo, R. B. Jackson Collection.*

H.C.Rwy. Co. #108 on the turntable of Waiakea round house, March 1923. *R. B. Jackson Collection.*

Front end view of H. C. Rwy. Co. #192 at the round house Waiakea 1946. *Cochrane Collection, Bishop Museum.*

Views of the Waiakea round house, Hawaiian Consolidated Rwy., near Hilo, Hawaii 1946. *Cochrane Collection, Bishop Museum.*

Mixed train of the H.C. Rwy. Co., crossing a high trestle, North Hilo coast, Island of Hawaii 1946. *Cochrane Collection, Bishop Museum.*

Hilo Railroad Co. #99 backs down on the Glenwood line with a mixed consist near the famed Volcano House about 1904. The track is so newly laid it is yet unballasted. *Camera of Hawaii.*

Hilo Railroad Co. #15. A Mason Machine Works locomotive working on the Hilo Breakwater about 1909. *Harry Wessel Collection.*

Hilo Railroad Co. train crossing the road near Waiakea and Hilo with #108 on the point. Date uncertain, probably about 1910. *Harry Wessel Collection.*

The new "Mountain View" station of the Hilo Railroad Co. about 1904. #99 on the point. Cane fields, native Japanese dress and elevated "Lanai" style homes vie with new rails and unballasted roadbed. *Camera of Hawaii.*

Short tunnel on the Hilo Railroad featuring lady driven rail car and overhead sugar cane flume. Lady's dress could place this photo at about 1910. *Bishop Museum.*

Deep cut on the Hilo Railroad, Island of Hawaii. Indicative of construction problems encountered in building the road. Date uncertain. *H.S.P.A. Library.*

Bridge construction across the Wailuku River near Hilo for the Hilo Rr. Co. The start of the expansion of the road toward Pauuilo about 1910. *Harry Wessel Collection.*

KAHULUI RAILROAD COMPANY
— KAHULUI, MAUI —

Original Kahului Railroad Company corporate seal, still in existence. A unique remnant of a once prosperous concern. *Alexander and Baldwin Inc.*

Later day Kahului Railroad token, circa 1935. Several thousand of these were stamped out for use by passengers on the railroad busline after rail passenger service was terminated. *Alexander and Baldwin Inc.*

(Kahului Railroad Co. records appear to be nonexistent because of an act of nature namely in tidal wave damage, yet old Hawaiian Newspapers have preserved much of the early scene by "Letters to the Editor". A primary source of such letters has been the "Hawaiian Gazette", especially in its issues of mid 1879 which follow here.)

The Editor, Hawaiian Gazette
Kahului, Maui July 3, 1879
Kahului & Wailuku Railroad
Dear Sir:
Captain Hobron had the good fortune to engage the services of Mr. E. Bailey Sr., of Wailuku to survey his railroad from Kahului to to Wailuku. Mr. Bailey is a scientific and practical engineer and is coming along finely with the work in his department. The Captain has also engaged Mr. Henry E. Carson, a practical locomotive engineer, of fifteen years experience both in the United States and South America, in the latter country under the late Henry Meigs.
Mr. Carson received the new locomotive in pieces on the beach on the 26th of June, and with the assistance of Mr. Levi Perkins as assistant engineer, under very unfavorable circumstances put it together, and on July 2nd at 4 o'clock pm, had steam up and ran up and down the track with the locomotive, (QUEEN EMMA) whistling as if with joy, that she was the first locomotive ever put in motion in these islands.
Your humble servant, of twenty years experience in railroading, superintends the construction of the roadbed, bridging, tracklaying, &c. The work is progressing very well with the inadequate force I have. The side track in the yard is laid, and the main track towards Makawao and one fourth mile towards Wailuku is also laid.
Respectfully,
J. McKinney

The Editor, Hawaiian Gazette
Kahului, Maui July 25, 1879
Dear Sir:
It might be interesting to your readers, or at least to yourself, to know of the progress of the Kahului & Wailuku Railroad. On the 30th ultimo the first stake was driven on the survey also ground broken and the first shovel full of earth thrown on the grade; on the 17th inst. the first section of the road was completed and an excursion train passed over it, and special trains are of daily occurrence since then. On the 21st inst. the first freight and accommodation train passed over the road, and is running regularly three trains of freight each way, and with a special train now and then, and the regular construction running, it makes times lively. Every thing seems to go off ship shape.
Respectfully,
J. McKinney

The Editor, Hawaiian Gazette
Kahului, Maui August 8, 1879
Dear Sir:
The work on the Kahului & Wailuku Railroad is being pushed forward with dispatch, and everything works lovely. The road is complete to Cape Horn, and every day the passenger and freight list increases. Today we have fifteen cars loaded each way with passengers and freight of the Wailuku, Waihee and Waikapu plantations, keeping the rolling stock on the road worked up to its full capacity, so I can scarcely get materials to the front for construction. You see by this that the little road is a success, and is worth more money than any plantation in the Hawaiian Islands. It shows the head of the enterprise has a head worth wearing. The force of men has been increased on construction and we hope to reach Wailuku soon; then we will drop you a line.
With Respect,
J. McKinney

Editor, H. G. Kahului, Maui September 16, 1879
I promised to let you know when we would reach Wailuku with the K. & W. Rr. We broke ground here on June 30th and commenced laying track on the first day of July and reached the depot ground in Wailuku on Sept. 10th at an expense of grading, bridging and tracking of about $600.00 per mile, and we are now ballasting and getting the track ready for the "Lightning Express" trains, which are to start in a week or so. The road has 15 curves in the entire length; the latest is at the depot at Kahului and is an eighteen degree curve; the next sharpest is at Cape Horn, and is only a seven degree curve and not at all detrimental to fast running.
There are two trestle bridges on the road — one 200 feet long and one 150 feet, one box culvert, one rock cut of 150 feet long and six feet deep. The heaviest grade on the road is 137 feet to the mile for about a half mile in length but the little engine seems to take its regular freight and passenger over the road with great ease.
There is already a heavy traffic over the road of both freight and passengers, Tuesdays, Wednesdays, Thursdays and Fridays; so I can have the train only on Saturdays and Monday of each week for hauling gravel for ballast; but there is another locomotive under way from the Baldwin Locomotive Works in Philadelphia, and material for more cars.

KAHULUI RAILROAD CO.

We will soon be up to the times with first class rolling stock in every respect, though as it is now, we pride ourselves with having the best there is in the Kingdom, and will play second to none, but run the best coaches and the fastest trains under management of steady, experienced employees, with whom the lives and property of the public will be in perfect safety.

With Respect,
J. McKinney

Editor, Hawaiian Gazette

Kahului, Maui October 10, 1879

Dear Sir:

All is peace and harmony about Kahului as usual; there is considerable shipping in port at present and the place seems destined to become a port of no little importance in its shipping and railroad enterprises. Captain Hobron is so well pleased with the success of his road to Wailuku that he has commenced and will soon have in operation another railroad from here to Hamakuapoko, work was commenced on it on the 6th inst.

The Captain seems to talk railroad now in earnest, and he is determined not to be outdone in the enterprise, not only on this island but any in the group. The citizens of Honolulu need not be surprised if they wake up some fine morning, at no distant day, by the whistle of Captain Hobron's locomotives as they pass through the street coming from all quarters of both the length and breadth of the Island of Oahu.

Captain Hobron knows which side of his bread is buttered, and he waits till the iron is hot and then strikes. We wish him success with the Kahului and Hamakuapoko Railroad.

With Respect,
J. McKinney

Hawaiian Gazette, April 17, 1880, an item—
Kahului Railroad, Maui
A Chinaman was killed by being run over by the cars at Kahului. Quite civilized up here!

In an obviously posed photo, Kahului, Wailuku & Hamakupoko Railroad Co., 3rd #1, the "Claus," sister engine to the first Baldwin Locomotive in the Hawaiian Kingdom, draws a "Kalakaua Commoner's Coach", on the Island of Maui. Circa 1886. *E. Burton Holmes 1901.*

K.Rr. Co. 3rd #1, the "Claus" at Kahului Harbor about 1910. *Archives of Hawaii.*

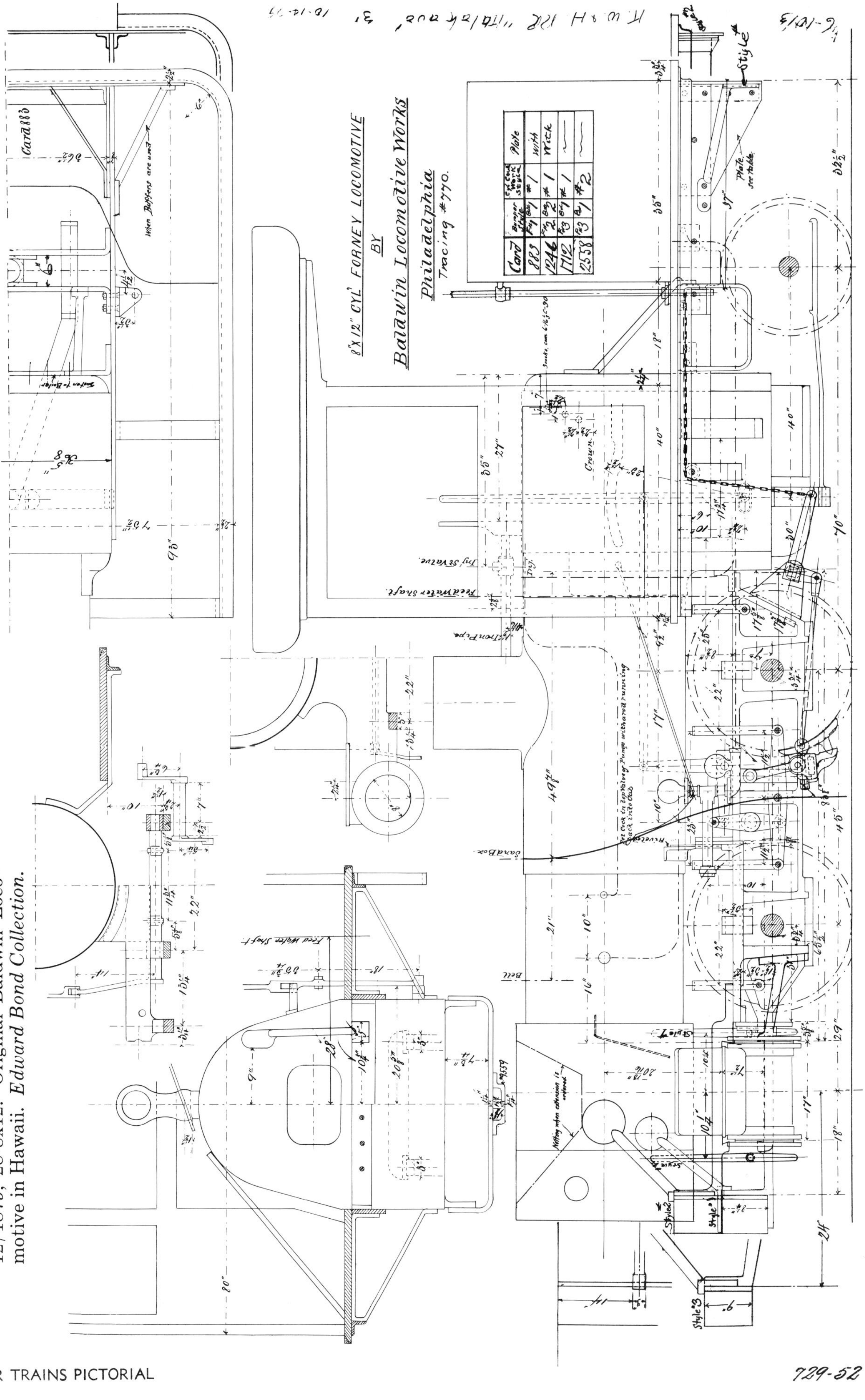

Locomotive "Kalakaua," Kahului Rr. Co. 2nd #1. Later "Leslie," 2nd #1 of the Hawaiian Railroad Company. Baldwin Class 6 10 1/3 C 7; Construction # 4906. Built 12/1879, 28-8x12. Original Baldwin Locomotive in Hawaii. *Edward Bond Collection.*

729-52

Two views labeled — "New York Elevated Railroad" — "Kalakaua" — the first Baldwin locomotive in the Hawaiian Kingdom, of the old Kahului Railroad, and later named the "Leslie" on the Hawaiian Rail-Road Co., draws a Kalakaua Coach at Rancho Rinconada, California, about 1965.

This coach, built by Ransomes & Rapier in England. drew King Kalakaua to the unveiling of the statue of King Kamehameha I, at Kohala in May of 1883. Front end view of "Kalakaua" at Rancho Rinconada about 1965. *Lindley Bothwell.*

Ex Hawaiian Railroad Co. #1 as K.Rr. Co. #2 on the old Pier #1 at Kahului Harbor about 1910. *Bishop Museum.*

First sugar produced at new Puunene mill in 1902 is hauled by the Kahului Railroad Co., whose main office appears in the background. *Allan Correll Collection.*

Connection of the Kahului Railroad main-line to the H.C. & S. Co. sugar mill. Circa 1905. *Allan Correll Collection.*

View of the Paia Depot, Paia, Maui, on the new Haiku extension of the Kahului Rail-road in 1905. Paia sugar factory can be seen near-by. *Allan Corell Collection.*

Kahului Railroad station, freight and post office about 1910. *Allan Corell Collection.*

Construction of the steel bridge over the Maliko Gulch on the Kahului Railroad on the Haiku extension in 1913. *Mr. Takayama, Allan Corell Collection.*

Old railyards, engine house and lumber yard of the Kahului Railroad about 1910. *Harry Wessel Collection.*

Pier #1 and rail installations at the railhead of the Kahului Railroad Co. about 1910. *Amfac Inc.*

Kahului Railroad Co. #10 moves loaded cars through the Kahului yards after 1946. *Alexander and Baldwin Inc.*

K.Rr. Co. #11 approaches the main line while switching in the Kahului yards sometime after 1941. *Alexander and Baldwin.*

Early view of K.Rr. Co. #3, "Mahulia" with a passenger train near Kahului Depot circa 1912. *Paradise of the Pacific.*

"Mahulia" hauls a train of bagged sugar from the Hawaiian Commercial & Sugar Company mill at Puunene to Kahului docks about 1915. *George M. Rolph.*

Kahului Rr. Co. #7 slumbers in the roundhouse at Kahului in May of 1949. *Clyde Couse.*

The "Paia Express" at the Wailuku Station of the Kahului Rr. Co. This H. K. Porter locomotive, the #4 machine of the railroad was later wrecked and after a short restoration period was scrapped. *John Allan.*

Kahului Rr. Co., road engine #5 in the livery of the Waialua Agricultural Company at Waialua, Oahu, in 1946. *Cochrane Collection, Bishop Museum.*

Builder's photo of K.Rr. Co. #7, BLW Construction #35892 of 1910. *G. M. Best Collection.*

K.Rr. Co. #8 on the old Kahului Railroad Pier #1 about 1915. *Bishop Museum.*

KAHULUI RAILROAD CO.

Originally numbered #6 this machine became road locomotive #8 at Kahului. Baldwin Class 10 8 ¼ D 11; Construction # 31125 Built 1907 40-12x18. *Edward Bond Collection.*

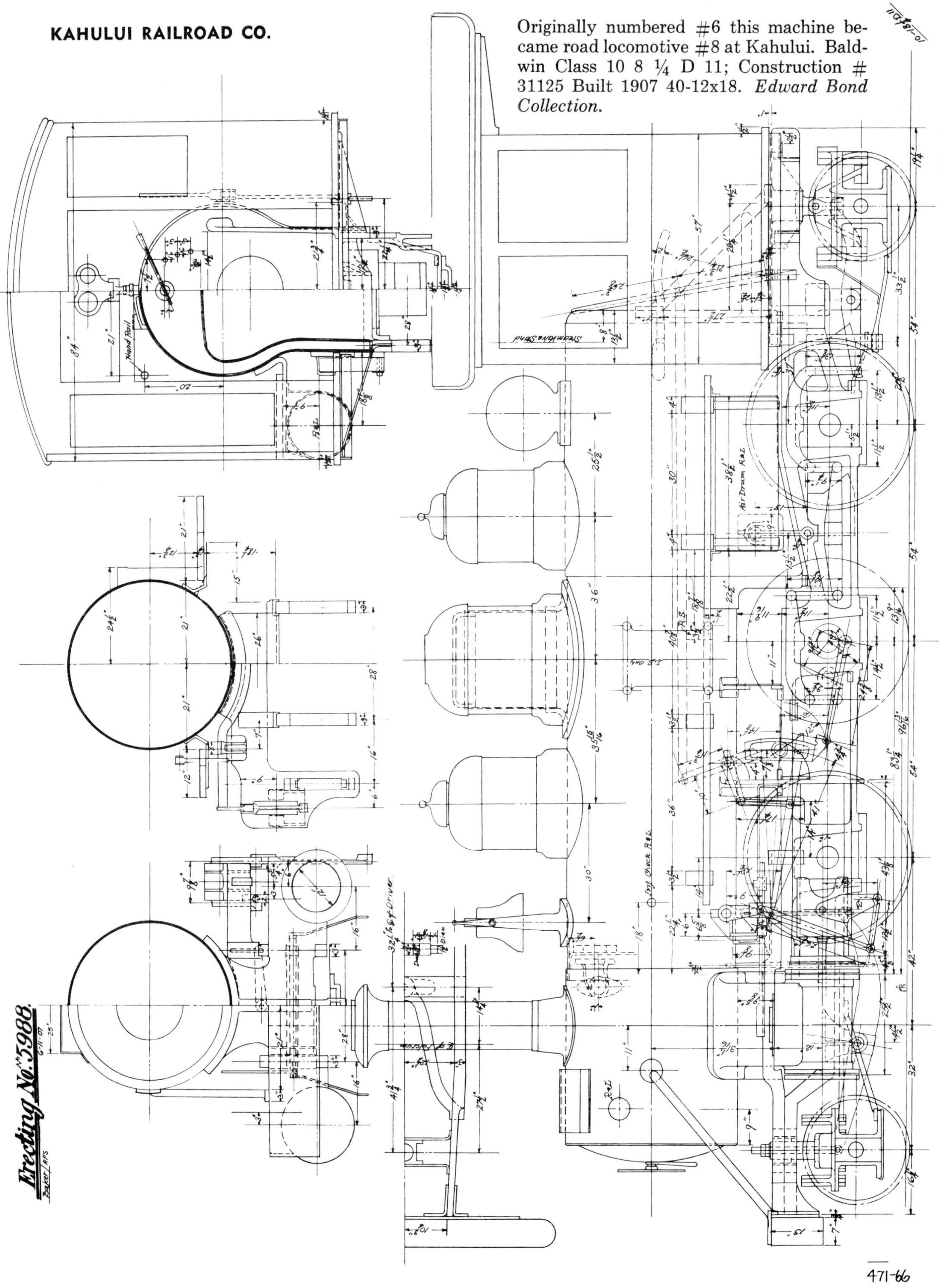

K.Rr.Co. #8 BLW Construction #31125 in the livery of the Waialua Agricultural Co. in the plantation yards on the Island of Oahu in 1946. *Cochrane Collection, Bishop Museum.*

Builder's photo of Kahului Rr. Co., road locomotive #9 of 1915. *G. M. Best Collection.*

K.Rr. Co. #9 on the rip track at Kahului in 1953. *G. M. Best Collection.*

Builder's photo of K.Rr. Co. #11. One of three sister machines of the same 2-6-2 or "Prairie" style as used in the later service of the railroad. *G. M. Best Collection.*

Perhaps the most photographed machine of the Kahului Railroad Co., and the last operating steam locomotive of that firm we see the #12 in many views on these two pages.

A.
K.Rr. #12 urges the morning consist into action at Kahului about 1960. *Murray Befler, Photo Hawaii.*

B.
Much used publicity shot of #12 in her heyday some 30 years after her debut at Kahului. *Murray Befler, Photo Hawaii.*

C.
#12 crosses the Maliko Gulch with a mixed consist about 1961. *Murray Befler, Photo Hawaii.*

D.
On a rail fan trip to Haiku #12 hustles along with its happy group, about 1964. *Murray Befler, Photo Hawaii.*

E.
Close-up of K.Rr. Co. #12 on one of its final rail-fan trips in 1964. *Elliott Whiton.*

A.

B.

C.

D.

E.

Oahu Railway & Land Company combine, built on Oahu in 1921 and later shipped to Maui, reposes in the rail yards at Kahului. In Kahului Rr. Co. livery in 1967 this car was later sold to "The Landing" restaurant on the edge of Kahului Harbor and converted there to a "Bar" car with clerestory top. *Lindley Bothwell.*

Early model diesel of the Kahului Railroad which was to prove sinister to the road in that most of the rail steam locomotives were supplanted by diesel motive power. *Robert Kemble.*

Kahului Railroad molasses cars made from flat cars and tanks in the Kahului yards in May of 1965. *Author's Collection.*

Covered sugar cars at Kahului Railroad in May of 1965. This style car was designed and constructed by the railroad for the local transport of bulk sugar. *Author's Collection.*

KAHULUI RAILROAD CO.

REVENUE STAMPS OF THE KAHULUI RAILROAD CO.

NOTE: *(Excerpted from an article by Russell O. Hill, in "Linn's Weekly Stamp News" of January 11, 1965 and used by permission of Mr. Hill.)*

While still under the control of the Wilder management a series of stamps were ordered by the Kahului Railroad Co. for its expanding merchandise business. The American Bank Note Co., of New York received the order on August 3, 1894 for a set of stamps, listed below, and which were delivered to the railroad on December 17, 1894. The denominations as shown below were printed in sheets of 100 stamps, cut into panes of 50 and shipped that way to the railroad company.

The adhesives were lithographed, the die being engraved on stone. These stamps were used by the Kahului Railroad Co. to prepay freight on packages, parcels and letters. They were attached to the package or letter itself and a duplicate receipt was given to the station agent to make up his way bills. The firm had the full cooperation of the Hawaiian postal service at that time, as the fast service the company rendered could not be supplied by the postal department. The stamps were printed on white wove paper perforated 12.

The quantities issued are recorded as —

5¢ blue	100,000
6¢ rose	100,000
15¢ green	75,000
18¢ black	100,000
$1.00 brown	50,000
50¢ purple	50,000

In addition to the American Bank Note company's issue there is also a set of these stamps done by another company. They also were lithographed, perforated 12, on white wove paper and were issued in denominations of 5¢ blue, 5¢ ultramarine, 10¢ carmine and 25¢ buff. They were produced by the Schmidt Lithographing Co., of San Francisco, California. They were used until late in 1898 or early 1899 when Hawaii was annexated to the United States.

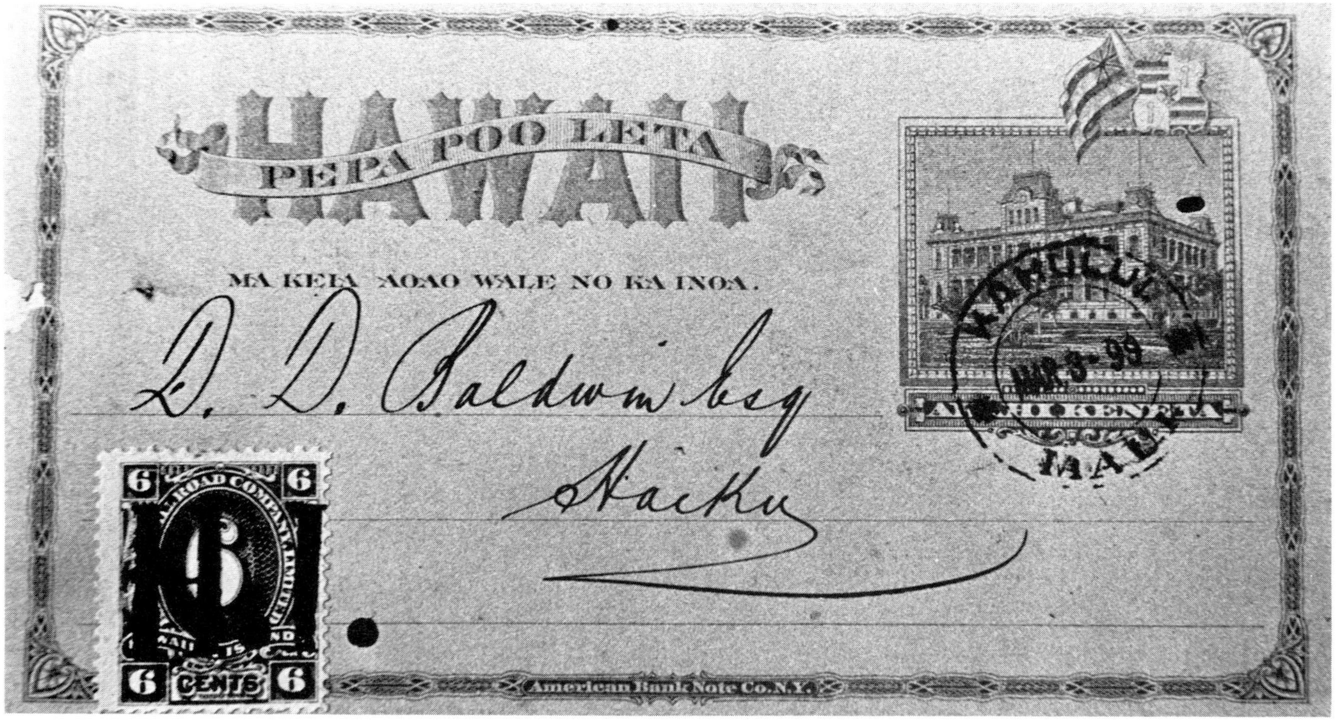

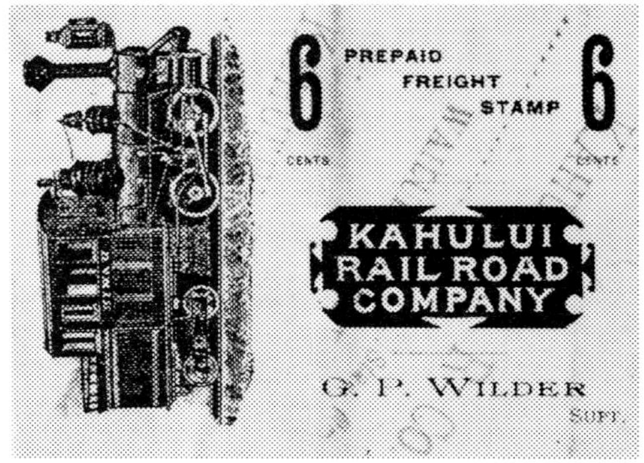

Facsimile of what is believed to be the original style and first issue of a Kahului Railroad revenue stamp. Its date of origin and useage is being investigated. *Colonel Pat Hogan, Stamps of Old Hawaii, Honolulu.*

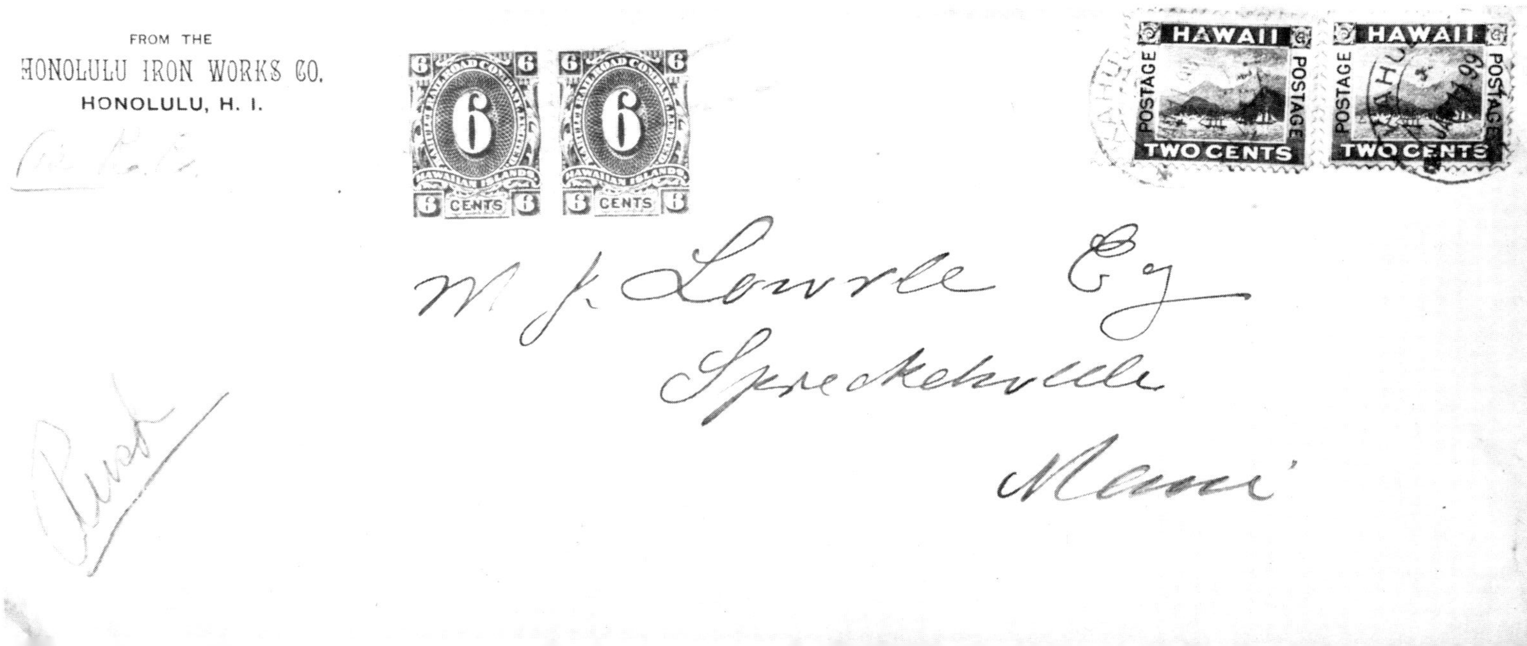

Facsimile of railroad and Hawaiian stamp covers. In use the stamps were cancelled with either red or blue pencil strokes, a hand stamped CANCELLED or a large H. I. stamped in purple ink.

KAUAI RAILWAY COMPANY
—PORT ALLEN, KAUAI—

Builder's photo of Kauai Railway Co. #1, "Port Allen." While this locomotive experienced several ownerships it is still in existence as "Wainiha" at the old Grove Farm Plantation. *Mallory H. Ferrell Collection.*

1900 builder's photo of "Wahiawa" of the McBryde Sugar Company of Kauai. This machine finally became the #2 road locomotive of the Kauai Rwy. Co. at Port Allen. Prior to its sale to the common carrier the McBryde firm removed the front pony truck and the tender. *Mallory H. Ferrell Collection.*

KOOLAU RAILWAY COMPANY
— HONOLULU, HAWAII —

Koolau Railway Company main line railroad as it travelled through a grove of Hau wood trees on Windward Oahu, near Hauula, in 1910. *Baker Collection, Bishop Museum.*

Koolau Rwy. Co. #1, in the livery of the Kahuku Plantation Co. at Kahuku on July 23, 1941. *Mallory Hope Ferrel, G. M. Best Collection.*

Kahuku Plantation Co. #3 on stand-by at the old Koolau Rwy. station near Kahuku in 1946. *Cochran Collection, Bishop Museum.*

Koolau Rwy. Co. #2 posing as Kahuku Plantation Co. 2nd #2 at Kahuku, shortly after being outshopped by the plantation. Date uncertain. *Henry F. Bonnell.*

OAHU RAILWAY & LAND COMPANY
— HONOLULU, OAHU —

A Severin photo of OR&L Co., "Lanakila" proudly posed by her crew in Hawaii. Her original name of "General Valleho" was changed to "Ben F. Dillingham" and her road number later became #45 in honor of Mr. Dillingham on his 45th birthday. *Baker Collection, Bishop Museum.*

A Baldwin veteran of 1875 vintage, BLW construction number 3732 works for the Hawaiian Dredging Co., in construction of the Oahu Railway. Circa 1889. *Baldwin Locomotives 1926*

Her active service complete the ancient "Ruby Hill Railroad" #1, (originally built for the Eureka & Consolidated Mining Co. of California in 1875) heads the OR&L rip track consist at a Waipahu siding in 1946. Her sister engines were OR&L Shays #43 and #44. *Cabral Collection.*

Below, Lanakila (Lit.: Victory), a Baldwin of 1881 vintage and originally the #3 locomotive of the Sonoma Valley Railroad was brought to Hawaii by the Barkentine "Planter" and works proudly on the road that was to be her home for many years. *Archives of Hawaii.*

Above: Built for the Park & Cliff House Railroad of San Francisco, OR&L Co. #5 (BLW #8973, 2-4-2-T) poses near the Honolulu Depot with its crew about 1905. It was reported as scrapped in 1923. *Harold Wilmunder*. Below: "Kauila" stands in the OR&L yards at Honolulu and is shown prior to its conversion to an 0-4-2. It was also the first locomotive to operate on the railroad's line. *Honolulu Advertiser photo.*

Above: OR&L Co. #6, an 0-4-0T Baldwin, was originally built for George A. Rowell of San Francisco. It was named Kauila (Lit.: Lightning) by the railroad and stands here on a railway pier at Honolulu about 1895. *Harold Wilmunder*. Below: OR&L Co. #6 on display near the Honolulu roundhouse on Fedruary 15, 1937 and before its present false stack and other items were appended. Still in existence it has become a part of the "Falls of Clyde" display on the docks of Honolulu. *Fred Stindt photo.*

Above: OR&L Co. #8 an 0-4-2T of 1878 vintage and named "Goliath" by the railroad, idles on a railway pier at Honolulu, circa 1895. Purchased from the Sonoma Valley Railroad of California it was named "Sonoma" in that service. *Harold Wilmunder*. Below: "Kaala" the #19 locomotive of the Oahu Railway appears to "step-out" for Waianae about 1900. *Baker Collection, Bishop Museum.*

Abov
Hawc
old O

A load of the "Best Sydney Coal" arrives at the railroad pier #17 in 1900. *Archives of Hawaii.*

Oahu Railway train of early vintage near a grass shack depot at Puuloa near the present area of Hickam Field. Some authorities believe the photo was taken from the lanai of the Damon residence once nearby. *R. B. Jackson Collection.*

Plantation and Common carriers dispute the "right-of-way" in 1890. The "Pacific Commercial Advertiser" placed this wreck for us on May 15, 1890, occurring at 1:00 p.m. the paper commented—"due to the carelessness of the OR&L Co." (Oh, My!) *Fred Denison Collection.*

An early view of veteran locomotive "Leahi" (Lit.: Diamond Head), OR&L Co. #15 as posed for a company photo about 1900. *Harold Wilmunder.*

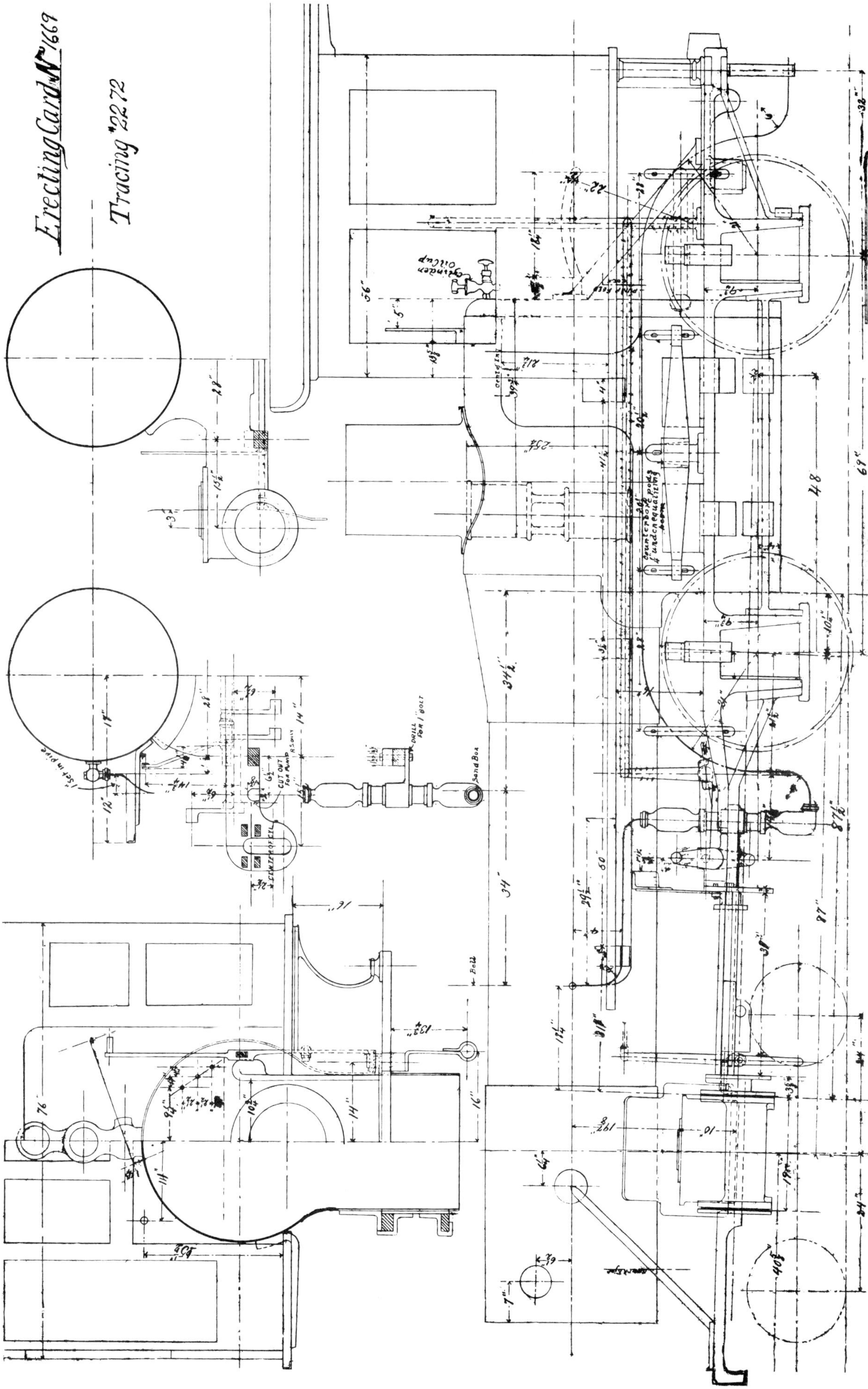

Locomotive "Leahi," OR&L Co. #15. Baldwin Class 8 11 C 1; Construction # 10129. Built 1889 33-9x12. *Edward Bond Collection.*

OR&L Co. officials, including Fred C. Smith try out a new "speeder" in 1906. *Fred Denison Collection.*

Fred C. Smith and Mr. Tarelton appear to express great satisfaction with the OR&L "speeder" of 1906. *Fred Denison Collection.*

Two views of the method of trestle building used by OR&L construction crews near Keanna Point about 1892. *Fred Denison Collection.*

For convenience of regular patrons of the Railway we issue COMMUTATION TICKETS, which are good for 50 Single Trips between points named on the face of the Ticket, and are good for the passage of the holder and his wife, actual members of his family, and his house servants. These Tickets are sold at

50 Per Cent. Reduction on Regular Single Trip Fares.

For convenience of Pearl City Lot owners we issue COMMUTATION TICKETS which are good for 50, 20, and 10 Single Trips.

Oahu Railway & Land Co.

TIME TABLE No. 11.

TO TAKE EFFECT

FEBRUARY 1st, 1896.

There are also 1000 Mile Tickets sold, in book form, which are good for 1000 miles travel over the road of the Company between ANY points. These tickets are good for passage of holder and his wife, actual members of his family and house servants. These Tickets are sold at

FIRST CLASS, ... $25.00
SECOND CLASS, ... 17.50

RATES OF FARE.

HONOLULU TO	1st Class Single	2nd Class Single	1st Class Round	2nd Class Round
Moanalua	.15	.10	.25	.15
Puuloa	.30	.20	.55	.35
Aiea	.40	.25	.70	.45
Kalauao	.45	.30	.80	.55
Waiau	.50	.35	.90	.65
Pearl City	.55	.40	1.00	.75
Waipio	.60	.45	1.10	.85
Waikele	.65	.50	1.20	.95
Hoaeae	.70	.55	1.30	1.00
Honouliuli	.75	.60	1.40	1.10
Ewa Mill	.85	.65	1.50	1.15
Waianae	1.35	1.10	2.20	1.70

Conductors collecting Train Fares must, in the presence of the passenger, paying such, punch a Ticket, distinctly stating whether same is FIRST or SECOND CLASS, with the name of the Station FROM and TO which the fare is paid, and deliver the sam eto the passenger.

No Round Trip Fares will be Collected on Trains.

Round Trip Tickets to Pearl City, Ewa Mill and Waianae are sold Saturdays and Sundays which are good between Saturday A. M. and Monday A. M. These tickets are sold as follows:

Pearl City....First Class, $ 75. Second Class, $ 50
Ewa Mill..... " " 1 00. " " 75
Waianae..... " " 1 50. " " 1 25

N. B.—Passengers not purchasing Tickets before boarding the train at stations where Tickets are sold will be charged 10 cents extra to the regular fare for TRAIN RATES.

Conductors are instructed and compelled to collect the same

No. 11. OAHU RAILWAY & LAND CO. Feb. 1, 1896.

FROM HONOLULU.							TOWARD HONOLULU.					
9	7	5	3	1				2	4	6	8	10
WAIANAE PASS.	WAIANAE PASS.	EWA MILL PASS.	WAIANAE PASS.	EWA MILL FREIGHT.	DISTANCE FROM HONOLULU.	STATIONS.	DISTANCE FROM WAIANAE	WAIANAE PASS.	EWA MILL FREIGHT.	WAIANAE PASS.	EWA MILL PASS.	WAIANAE PASS.
Sat.&Sun. Only.	DAILY	Daily, Ex. Sat.&Sun.	DAILY.	DAILY. EX. SUN.				DAILY.	DAILY. EX. SUN.	DAILY.	Daily, Ex. Sat.&Sun.	Sat.&Sun. Only.
P. M.–LV.	P. M.–LV.	P. M.–LV.	A. M.–LV.	A. M.–LV.				A. M.–AR.	A. M.–AR.	P. M.–AR.	P. M.–AR.	P. M.–AR.
1.45	5.10	1.45	9.15	6.40	0.0	HONOLULU	33.3	8.23	10.30	3.11	4.55	5.55
1.53	5.18	1.53	9.23	....	2.7	MOANALUA	30.6	8.15	....	3.03	4.47	5.47
2.01	*5.26	2.01	9.31	....	6.1	PUULOA	27.2	8.07	....	2.55	4.39	5.39
2.09	5.34	2.09	9.39	....	8.7	AIEA	24.6	7.59	....	2.47	4.31	*5.31
2.12	5.37	2.12	9.42	....	10.0	KALAUAO	23.3	7.56	....	2.44	4.28	5.28
2.15	5.40	2.15	9.45	....	11.0	WAIAU	22.3	7.53	....	2.41	4.25	5.25
2.18 / 2.28	5.43 / 5.53	2.18 / 2.28	9.48 / 9.58	7.25 / 7.40	11.7	PEARL CITY	21.6	7.50 / 7.40	9.48 / 9.38	2.38 / 2.28	4.22 / 4.12	5.22 / 5.12
2.32	5.57	2.32	10.02	....	13.6	WAIPIO	19.7	7.36	....	2.24	4.08	5.08
2.35	6.00	2.35	10.05	....	14.6	WAIKELE	18.7	7.33	....	2.21	4.05	5.05
2.38	6.03	2.38	10.08	....	15.2	HOAEAE	18.1	7.30	....	2.18	4.02	5.02
2.43	6.08	2.43	10.13	....	16.3	HONOULIULI	17.0	7.25	....	2.13	3.57	4.57
2.49	6.14	2.49	10.19	8.10	18.3	EWA MILL	15.0	7.19	9.10	2.07	3.51	4.51
3.24	6.49	P. M.–AR.	10.54	A. M.–AR.	33.3	WAIANAE	00.0	6.44	A. M.–LV.	1.32	P. M.–LV.	4.16
P. M.–AR.	P. M.–AR.		A. M.–AR.					A. M.–LV.		P. M.–LV.		P. M.–LV.

* Trains No. 7 and No. 10 will meet at Halawa Siding at 5:32 p. m.
Freight Trains will carry Passenger Accommodations.

B. F. DILLINGHAM,
General Manager.

G. P. DENISON,
Superintendent.

Oahu Railway Timetable of 1896. *Dillingham Corporation.*

1903 Timetable for the railroad. May 1903. *Dillingham Corporation.*

Oahu Railway and Land Co.

Time Table

May 1, 1903

B. F. Dillingham
General Manager

Geo. P. Denison
Superintendent

F. C. Smith
Gen'l Pass. and Ticket Agent

THE MERCANTILE PTG. CO. LTD.

RATES OF FARE

HONOLULU TO	FIRST CLASS Single	SECOND CLASS Single	FIRST CLASS Round	SECOND CLASS Round
Moanalua	10	10	15	15
Puuloa	25	20	45	35
Aiea	30	25	55	45
Kalauao	35	30	65	55
Waiau	40	35	70	60
Pearl City	45	40	80	70
Waipio	50	45	90	80
Waipahu	50	45	90	80
Hoaeae	55	50	1 00	85
Honouliuli	60	55	1 10	90
Ewa Mill	70	60	1 20	1 00
Waianae	1 15	1 00	1 90	1 60
Makaha	1 20	1 05	2 00	1 65
Makua	1 45	1 20	2 35	1 90
Kawaihapai	1 80	1 45	2 90	2 35
Mokuleia	1 85	1 50	3 00	2 45
Puuiki	1 95	1 55	3 15	2 55
Waialua	2 00	1 60	3 20	2 60
Waimea	2 20	1 75	3 50	2 85
Kahuku	2 50	2 00	4 00	3 25

Passengers not purchasing tickets before boarding the train at Stations where tickets are sold will be charged 10 cents extra to the regular fare for Train Rates.

No Round Trip Fares will be collected on trains.

Children, five years of age and under twelve, half fare; under five years, when in charge of a competent person, free. Half fare will not be issued on Saturday to Monday Tickets.

Stop over privileges will be allowed upon application to the Conductor.

EXCURSION TICKETS

Excursion Tickets are sold on Saturdays and Sundays, and are good from Saturday Morning until the following Monday at Noon. These Tickets are sold as follows:

Pearl City First-Class, $ .65 Second-Class, $ .50
Waipahu " " .75 " " .60
Ewa Mill " " 1.00 " " .75
Waianae " " 1.50 " " 1.25
Waialua " " 2.00 " " 1.75
Kahuku " " 2.50 " " 2.25

PARLOR CARS

The following extra charge is made for seats in these cars:
Honolulu to Waianae....................25 cents
Honolulu to points beyond Waianae..............50 cents

TOWARD HONOLULU

STATIONS	Distance	2	18	4	20	6	22	12	16
		Ex. Sun.	Ex. Sun.	Daily	Daily	Daily	Daily	Daily	Daily
		A. M.	A. M.	A. M.	A. M.	P. M.	P. M.	P. M.	P. M.
Kahuku	0.0			5 35				2 13	
Waimea	10.6			5 59				2 38	
Waialua	15.5			6 10				2 55	
Puuiki	16.8			6 13				2 58	
Mokuleia	19.6			6 19				3 04	
Kawaihapai	21.0			6 23				3 09	
Makua	30.8			6 48				3 37	
Makaha	36.3			7 01				3 51	
Waianae	38.0			7 10				4 00	
Nanakuli	44.1			7 24				4 14	
Gilbert	48.8			7 35				4 25	
Ewa Mill	53.0	5 55	6 50	7 45	9 43	1 10	3 33	4 37	6 45
Honouliuli	55.0	6 01	6 56	7 50	9 49	1 16	3 39	4 43	6 51
Hoaeae	56.1	6 05	7 00	7 53	9 53	1 20	3 43	4 47	6 55
Waipahu	57.1	6 08	7 03	7 56	9 56	1 23	3 46	4 50	6 58
Waipio	57.7	6 10	7 05	7 58	9 58	1 25	3 48	4 52	7 00
Pearl City	59.6	6 16	7 11	8 03	10 04	1 31	3 54	4 57	7 06
Waiau	60.3	6 19	7 14	8 05	10 07	1 34	4 00	5 00	7 09
Kalauao	61.3	6 22	7 17	8 08	10 10	1 37	4 03	5 03	7 12
Aiea	62.6	6 26	7 21	8 12	10 14	1 41	4 07	5 07	7 16
Puuloa	65.2	6 34	7 29	8 20	10 22	1 49	4 15	5 15	7 24
Moanalua	68.6	6 42	7 38	8 28	10 30	1 57	4 23	5 23	7 32
Honolulu	71.3	6 50	7 46	8 36	10 38	2 05	4 31	5 31	7 40
		A. M.	A. M.	A. M.	A. M.	P. M.	P. M.	P. M.	P. M.

TOWARD KAHUKU

STATIONS	Distance	17	3	7	19	9	21	11	23	15
		Ex. Sun.	Daily	Daily	Daily	Daily	Ex. Sun.	Daily	Sun. only	Ex. Sun.
		A. M.	A. M.	A. M.	P. M.	P. M.	P. M.	P. M.	P. M.	P. M.
Honolulu	0.0	7 30	9 15	11 05	2 15	3 20	4 15	5 15	9 30	11 15
Moanalua	2.7	7 38	9 23	11 13	2 23	3 28	4 23	5 23	9 38	11 23
Puuloa	6.1	7 46	9 31	11 21	2 31	3 36	4 31	5 31	9 46	11 31
Aiea	8.7	7 54	9 39	11 29	2 39	3 44	4 39	5 39	9 54	11 39
Kalauao	10.0	7 58	9 43	11 33	2 43	3 48	4 43	5 43	9 58	11 43
Waiau	11.0	8 01	9 46	11 36	2 46	3 51	4 46	5 46	10 01	11 46
Pearl City	11.7	8 05	9 49	11 40	2 50	3 54	4 57	5 50	10 05	11 50
Waipio	13.6	8 10	9 54	11 45	2 55	3 59	5 02	5 55	10 10	11 55
Waipahu	14.2	8 12	9 56	11 47	2 57	4 01	5 04	5 57	10 12	11 57
Hoaeae	15.2	8 15	9 59	11 50	3 00	4 04	5 07	6 00	10 15	12 00
Honouliuli	16.3	8 19	10 03	11 54	3 04	4 08	5 11	6 04	10 19	12 04
Ewa Mill	18.3	8 25	10 09	12 00	3 10	4 14	5 17	6 10	10 25	12 10
Gilbert	22.5		10 19			4 25				
Nanakuli	27.2		10 30			4 35				
Waianae	33.3		10 50			4 54				
Makaha	35.0		10 54			4 58				
Makua	40.5		11 08			5 11				
Kawaihapai	50.3		11 36			5 36				
Mokuleia	51.7		11 41			5 40				
Puuiki	54.5		11 47			5 46				
Waialua	55.8		11 55			5 50				
Waimea	60.7		12 07			6 01				
Kahuku	71.3		12 32			6 25				
		A. M.	P. M.	M.	P. M.	P. M.	P. M.	P. M.	P. M.	A. M.

OR&L Logos of Note:

A.

C.

D.

B.

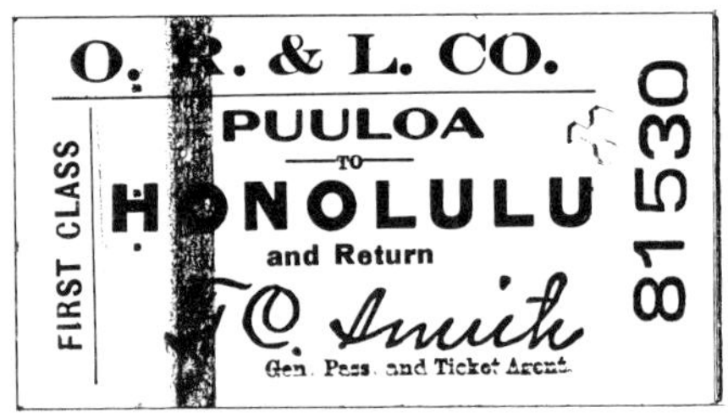

E.

A. Design used on freight cars circa 1906 and for Train uniform buttons. B. Painted in yellow on Tuscan Red painted Box Cars these designs advertised the local railroad. C. This design was a logo of service use and was reprinted in the Honolulu Star Bulletin of January 11, 1964 long after the demise of the old rail system. D. OR&L Loggo of internal use, and E. the First Class Ticket design. *Author's Collection, Courtesy Dilco.*

Two shy Hawaiian lads face the camera to assist in the display of an OR&L Co. Logo which once adorned OR&L boxcars. Circa 1906. *Fred Denison Collection.*

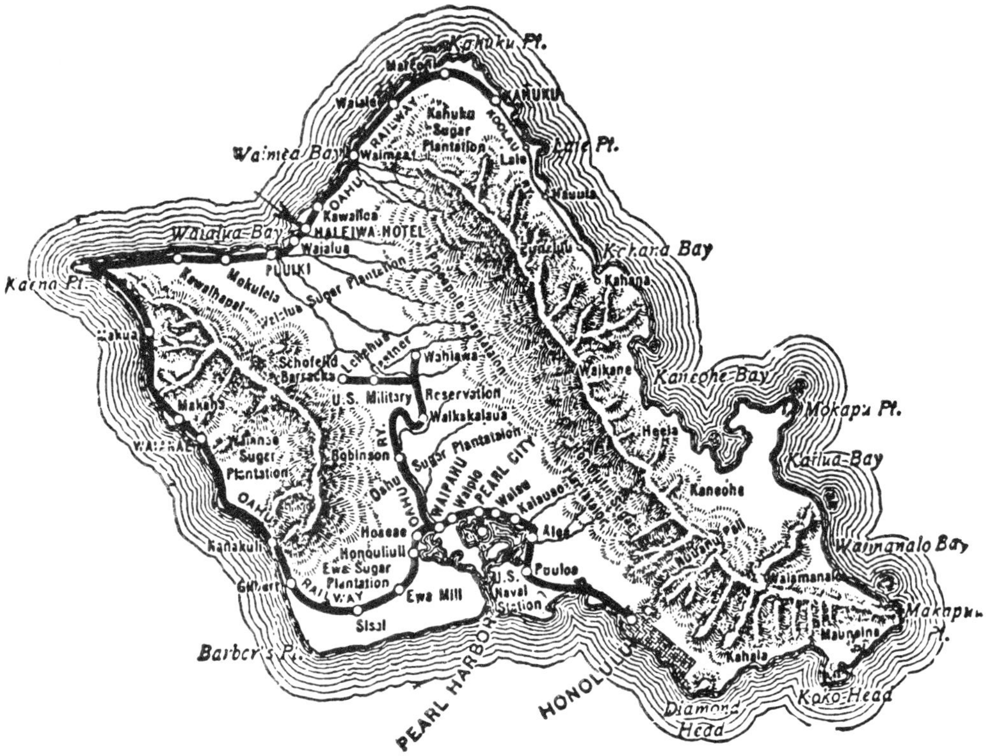

Photo map of the Cochrane Collection showing extent of the Oahu Railway line. *Cochrane Collection, Bishop Museum.*

"Kauila" as OR&L Co. #6 as restored with a false stack in the rail yards in 1946. *Cochrane Collection, Bishop Museum.*

Oahu Railway Depot circa 1946. *Cochrane Collection, Bishop Museum.*

Views of the rear of the Oahu Railway Co. Depot 1946. Station platform and Freight transfer shed. *Cochrane Collection, Bishop Museum.*

Builder's photo of OR&L #9 an 0-6-0 of 1910 vintage as built by Alco-Rodgers. *Alco Historic Photos Mohawk & Hudson Chapter NRS.*

Switcher #10 in OR&L yards at Honolulu 1946. *Cochrane Collection, Bishop Museum.*

OR&L #9 on standby assignment in Honolulu Yard 1945. (Note: Original photo cropped at top.) *Cochrane Collection, Bishop Museum.*

Oahu Railway switcher #12 in Honolulu service yards 1946. Still in existence this locomotive was being restored to steam service in 1974. *Cochrane Collection, Bishop Museum.*

"Leahi" as OR&L #15 at Waipahu on March 23, 1923. *R. B. Jackson Collection.*

Train to Waipahu with OR-&L "Leahi" on the point. *R. B. Jackson Collection.*

The "Waianae Express" with "Leahi" and car #34 passing through Waipahu in 1923. *R. B. Jackson Collection.*

Builder's photo of OR&L Co. #22. A Baldwin Vauclain Compound (BLW #20324—1902), its tremendous cylinder power created many breakdowns and cracked her frames. As a result, the engine was simpled by the OR-&L and the Honolulu Iron Works. *Gerald M. Best Collection.*

An in-service view of OR&L #22 after being simpled by the road. Once this machine ran away "down hill" and piled up at Quarry killing her engineer and breaking her frame. This accident hastened the decision to simple the engine. *Cochrane Collection, Bishop Museum.*

Road locomotive #31 prepares for her daily stindt at the service facility of the railroad in 1946. *Cochrane Collection, Bishop Museum.*

OR&L Co. #31 sister engine of OR&L #22 was originally also a Baldwin Vauclain Compound locomotive simpled by the railroad. (BLW #20325—1902). Shown here in the Honolulu yards in 1946. *Cochrane Collection, Bishop Museum.*

Close-up of OR&L "Kaala," road locomotive #19 of the OR&L Co., at Waipahu in 1923. *R. B. Jackson Collection.*

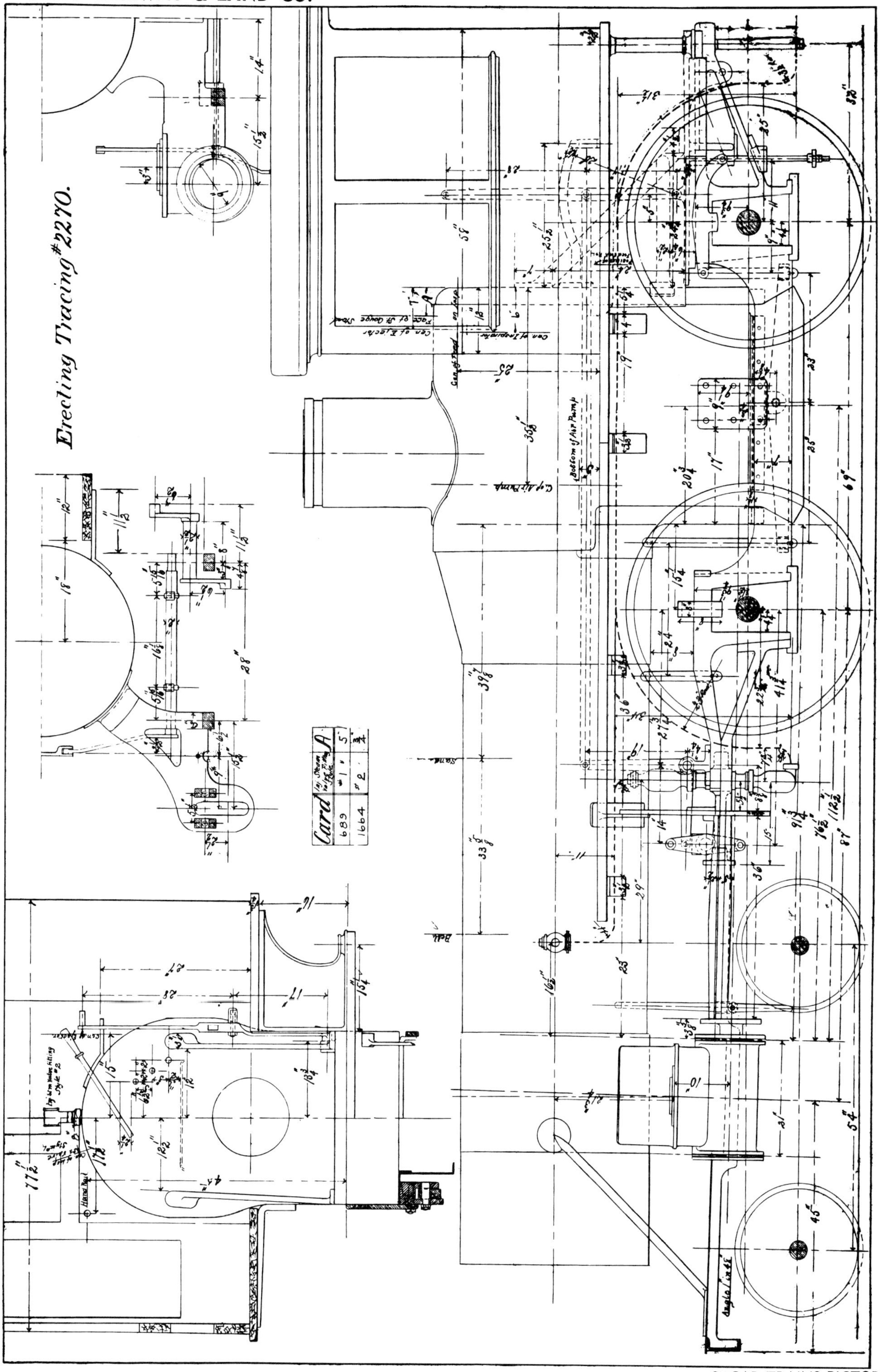

Locomotive "Kaala," OR&L Co. #19. Baldwin Class 8 12 C 6; Construction # 10131. Built 1889 37-9x16. *Edward Bond Collection.*

FORM 1007

Shop Nos. 46716.
Road Nos. 10.
Shipment Dec 15-1909

AMERICAN LOCOMOTIVE COMPANY.

Pittsburgh WORKS. SPECIFICATION. TYPE 0-6-0.

Copy sent to NY Office 1/13/10 — MADE BY — CHECKED BY — ORDER P-618. Drawing No. 438-S-2680 ROAD Oahu Ry & Land Co.

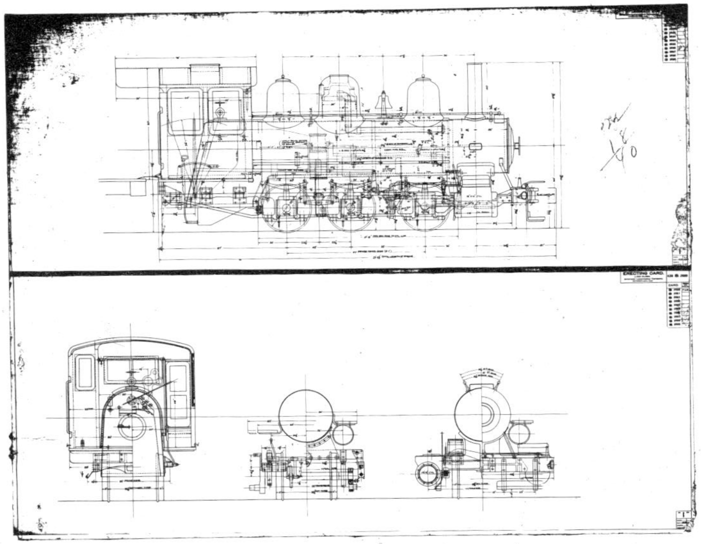

GENERAL DIMENSIONS.		
Gauge	3'-0"	
Fuel	Oil	
Weight on drivers in Running Order	46100	
Weight on Truck	—	
" " Trailers	—	
Weight of Engine in Running Order (in Red)	46100	
" of Tender in Running Order	~~16300~~	41900 w/½ (L'E'n'A)
" of Eng. and Tend. in Running Order	~~62400~~	88000
" of Engine Empty	41200	
" of Tender Empty	25750	
" of Engine and Tender Empty	66950	
Wheel Base, Driving	[illegible]	8'-1"
" " Total Engine	[illegible]	8'-1"
" " Total Engine and Tender	32'-10¾"	
Tractive Power, Maximum	9495	
Adhesion (Factor of)	4.21	
CYLINDERS.		
Diameter and Stroke	12"x16"	
Diameter of Piston Rod	2¼"	
Style of Piston Packing	Cast iron	
Steam Ports, Size	1"x10"	
Exhaust Ports, Size	2¼"x10"	
Crosshead, Type	Single Bar	
VALVES.		
Type of Gear	Stephenson	
Diameter (if piston)	—	
Greatest Travel	3½	
Outside Lap	⅝"	
Inside Lap ~~or Clearance~~	Line & Line	
Lead in Full Gear		
DRIVING WHEELS, ETC.		
Diameter Driving Wheels, Outside Tire	~~[illegible]~~ 33"	
" " " Centres	28"	
Tire Held by	Shrinkage	
Diameter Trailing Wheels, Outside Tire	—	
Driving Journals, Diameter and Length	5½"x6½"	
Trailing " " "	—	
Diameter of Driving Wheel Fits	5¾"	
Diam. and Length of Main Crank Pin Journals	3½"x3½"	
" " " Side Rod Journals	4"x3½"	
" " " F. & B. Journals	3"x2½"	
" " " Inter. Journals	—	
Engine Truck, Type	—	
" " Journals	—	
" " Wheels, Diameter	—	
Frames, Width	3"	
BOILER.		
Style	Straight Top	
Outside Diameter of First Ring	38"	
" " Largest Course	38 11/16"	
Height over Crown, Front	13 13/32"	
Working Pressure	160#	
Height to Center, From Rail	5'-0"	
Thickness of Plates, Shell	5/16" & 11/32"	

Dome, Diameter		24"
Fire Box Length	43 3/16"	~~[illegible]~~
" " Width		38⅜"
" " Depth		48"
" " Depth (top of grate to cen. lowest tube)		—
" " Plates Thickness, Sides		5/16"
" " " " Back		5/16"
" " " " Crown		⅜"
" " " " Tube Sheet		½"
" " Water Space, Front		3"
" " " " Sides		2½"
" " " " Back		2½"
" " Crown Staying		13/16 Radial
" " Stay Bolts		⅞"x1"
Tubes, Material		Charcoal Iron
" Thickness		#13
" Number of		67
" Spacing of		[illegible] 1⅝"
" Diameter		1¾"
" Length over Tube Sheets		10'-7"
Fire Brick, Supported on		Arranged for Oil Burning
Ash Pan		" " "
" Air or Steam Operated		" " "
Fire Door		14"
Heating Surface, Tubes		322. SQ.FT.
" " Water Tubes		—
" " Fire Box		58.
" " Total		380. " "
Grate Surface		11.4 "
" Style		Oil Burning
Exhaust Pipes		Single
" Nozzles		3"-3¼"
Smoke Stack, Diameter Inside		9¼"
" " Height Top Above Rail		9'-11 5/16"
Boiler Supported by		Expansion Pads
Cab Material		Steel
Brakes, Back or Front of Driver		Back
TENDER.		
Wheels, Number		8
" Diameter		24"
Journals, Diameter and Length		3¼"x7"
Wheel Base		11'-9"
Tender Frame		8" Steel Channel
Trucks, Style		Arch Bar type
Brakes, Inside or Outside Hung		Inside
Tank, Kind		U-Sloping
" Body, Length		15'-0"
" " Width		7'-4"
" " Depth		2'-6"
" Capacity, Gallons	1200 U.S. Gallons	
Coal Capacity, Tons		840 Gal oil
Length over all, Engine and Tender		41'-10¾"
Extreme Width		8'-4"
" Height		9'-11 5/16"

ANY UNUSUAL ITEMS OF DESIGN OR CONSTRUCTION APPLYING TO THIS ENGINE SHOULD BE NOTED ON FORM.

Specification sheet for OR&L Co. #10 switcher as built by Alco-Pittsburgh Works 1909. *Charles Purington.*

0-6-0 switcher #12 stares at the camera of Kent Cochrane from the service area of the Honolulu yards in 1946. *Cochrane Collection, Bishop Museum.*

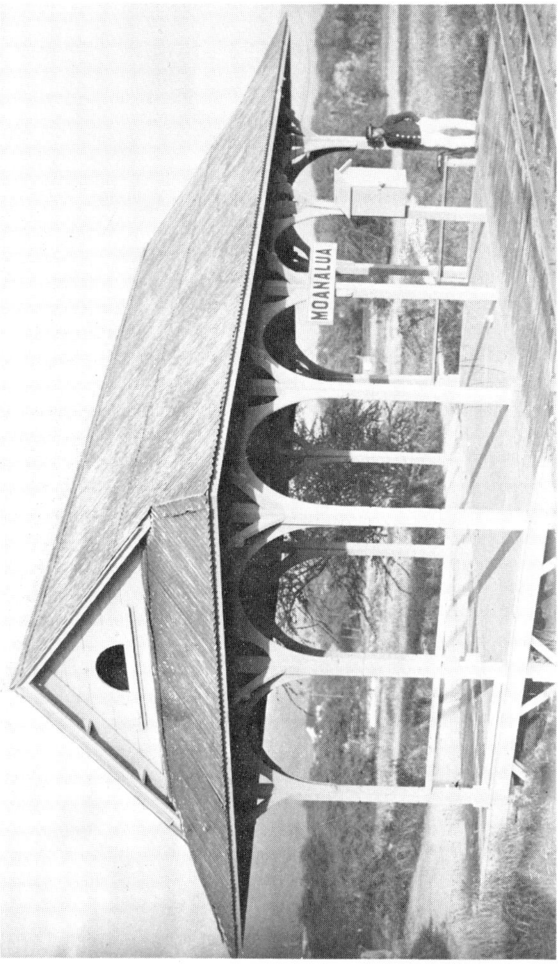

Right, above: Moanalua Station shelter of about 1906, omitted from later timetables this station may have originally been just a flag stop. Note Conductor's phone box. *Fred Denison Collection, H.S.P.A.* Right, below: The Puuloa station of the OR&L near Pearl Harbor entertains a solitary train rider. Date uncertain. *"Ace" Rowell Collection.*

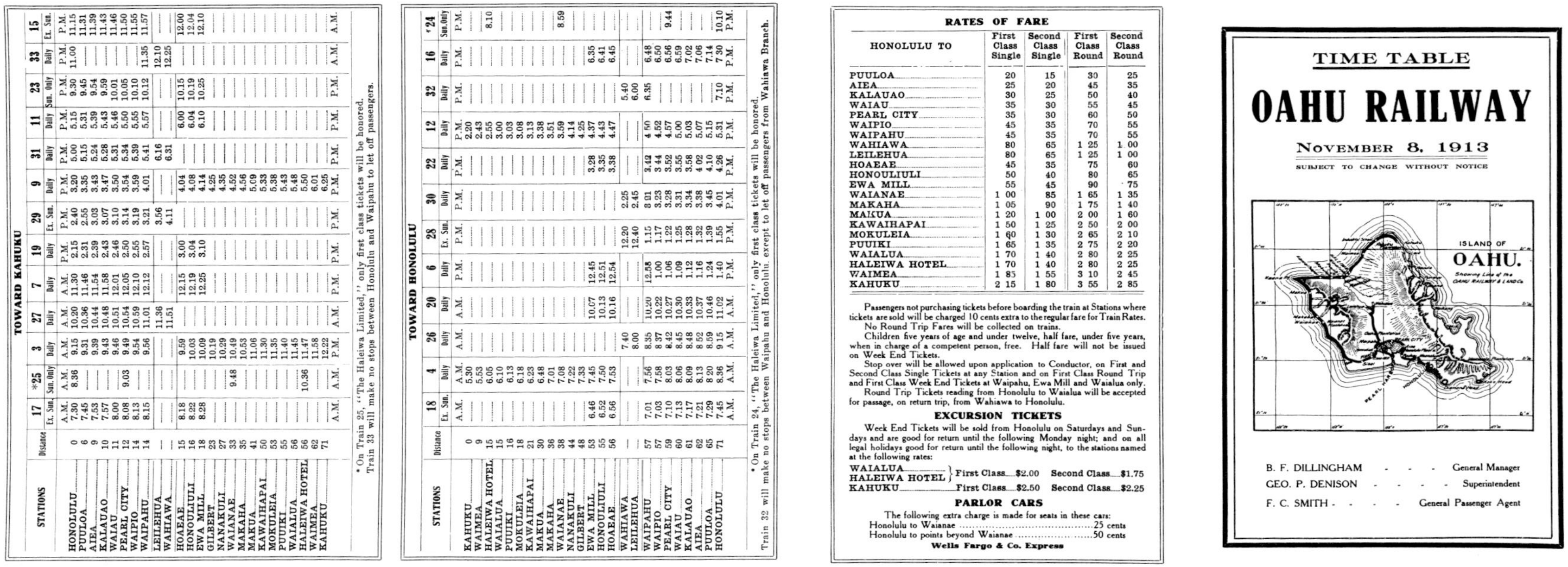

TIME TABLE

OAHU RAILWAY

NOVEMBER 8, 1913

SUBJECT TO CHANGE WITHOUT NOTICE

B. F. DILLINGHAM - - - - General Manager
GEO. P. DENISON - - - - Superintendent
F. C. SMITH - - - - General Passenger Agent

RATES OF FARE

HONOLULU TO	First Class Single	Second Class Single	First Class Round	Second Class Round
PUULOA	20	15	30	25
AIEA	25	20	45	35
KALAUAO	30	25	50	40
WAIAU	35	30	55	45
PEARL CITY	35	30	60	50
WAIPIO	45	35	70	55
WAIPAHU	45	35	70	55
WAHIAWA	80	65	1 25	1 00
LEILEHUA	80	65	1 25	1 00
HOAEAE	45	35	75	60
HONOULIULI	50	40	80	65
EWA MILL	55	45	90	75
WAIANAE	1 00	85	1 65	1 35
MAKAHA	1 05	90	1 75	1 40
MAKUA	1 20	1 00	2 00	1 60
KAWAIHAPAI	1 50	1 25	2 50	2 00
MOKULEIA	1 60	1 30	2 65	2 10
PUUIKI	1 65	1 35	2 75	2 20
WAIALUA	1 70	1 40	2 80	2 25
HALEIWA HOTEL	1 70	1 40	2 80	2 25
WAIMEA	1 85	1 55	3 10	2 45
KAHUKU	2 15	1 80	3 55	2 85

Passengers not purchasing tickets before boarding the train at Stations where tickets are sold will be charged 10 cents extra to the regular fare for Train Rates.

No Round Trip Fares will be collected on trains.

Children five years of age and under twelve, half fare, under five years, when in charge of a competent person, free. Half fare will not be issued on Week End Tickets.

Stop over will be allowed upon application to Conductor, on First and Second Class Single Tickets at any Station and on First Class Round Trip and First Class Week End Tickets at Waipahu, Ewa Mill and Waialua only.

Round Trip Tickets reading from Honolulu to Waialua will be accepted for passage, on return trip, from Wahiawa to Honolulu.

EXCURSION TICKETS

Week End Tickets will be sold from Honolulu on Saturdays and Sundays and are good for return until the following Monday night; and on all legal holidays good for return until the following night, to the stations named at the following rates:

WAIALUA, HALEIWA HOTEL } First Class $2.00 Second Class $1.75
KAHUKU First Class $2.50 Second Class $2.25

PARLOR CARS

The following extra charge is made for seats in these cars:
Honolulu to Waianae .... 25 cents
Honolulu to points beyond Waianae .... 50 cents

Wells Fargo & Co. Express

TOWARD KAHUKU

STATIONS	Distance	17	*25	3	27	7	19	29	9	31	11	23	33	15
		Ex. Sun.	Sun. Only	Daily	Daily	Daily	Daily	Ex. Sun.	Daily	Daily	Daily	Sun. Only	Daily	Ex. Sun.
		A.M.	A.M.	A.M.	A.M.	A.M.	P.M.	P.M.	P.M.	P.M.	P.M.	P.M.	P.M.	P.M.
HONOLULU	0	7.30	8.36	9.15	10.20	11.30	2.15	2.40	3.20	5.00	5.15	9.30	11.00	11.15
PUULOA	6	7.45		9.31	10.36	11.46	2.31	2.55	3.35	5.15	5.31	9.45		11.31
AIEA	9	7.53		9.39	10.44	11.54	2.39	3.03	3.43	5.24	5.39	9.54		11.39
KALAUAO	10	7.57		9.43	10.48	11.58	2.43	3.07	3.47	5.28	5.43	9.59		11.43
WAIAU	11	8.00		9.46	10.51	12.01	2.46	3.10	3.50	5.31	5.46	10.01		11.46
PEARL CITY	12	8.08	9.03	9.49	10.54	12.05	2.50	3.14	3.54	5.34	5.50	10.05		11.50
WAIPIO	14	8.13		9.54	10.59	12.10	2.55	3.19	3.59	5.39	5.55	10.10		11.55
WAIPAHU	14	8.15		9.56	11.01	12.12	2.57	3.21	4.01	5.41	5.57	10.12	11.35	11.57
LEILEHUA	—				11.36			3.56		6.16			12.10	
WAHIAWA	—				11.51			4.11		6.31			12.25	
HOAEAE	15	8.18		9.59		12.15	3.00		4.04		6.00	10.15		12.00
HONOULIULI	16	8.22		10.03		12.19	3.04		4.08		6.04	10.19		12.04
EWA MILL	18	8.28		10.09		12.25	3.10		4.14		6.10	10.25		12.10
GILBERT	23			10.19					4.25					
NANAKULI	27			10.29					4.35					
WAIANAE	33		9.48	10.49					4.52					
MAKAHA	35			10.53					4.56					
MAKUA	41			11.06					5.09					
KAWAIHAPAI	50			11.30					5.33					
MOKULEIA	53			11.35					5.38					
PUUIKI	55			11.40					5.43					
WAIALUA	56			11.45					5.48					
HALEIWA HOTEL	56		10.36	11.47					5.50					
WAIMEA	62			11.58					6.01					
KAHUKU	71			12.22					6.25					
		A.M.	A.M.	P.M.	A.M.	P.M.	P.M.	P.M.	P.M.	P.M.	P.M.	P.M.	A.M.	A.M.

* On Train 25, "The Haleiwa Limited," only first class tickets will be honored.
Train 33 will make no stops between Honolulu and Waipahu to let off passengers.

TOWARD HONOLULU

STATIONS	Distance	18	4	26	20	6	28	30	22	12	32	16	*24
		Ex. Sun.	Daily	Daily	Daily	Daily	Ex. Sun.	Daily	Daily	Daily	Daily	Daily	Sun. Only
		A.M.	A.M.	A.M.	A.M.	P.M.	P.M.	P.M.	P.M.	P.M.	P.M.	P.M.	P.M.
KAHUKU	0		5.30							2.20			
WAIMEA	9		5.53							2.43			
HALEIWA HOTEL	15		6.05							2.55			8.10
WAIALUA	15		6.10							3.00			
PUUIKI	16		6.13							3.03			
MOKULEIA	18		6.18							3.08			
KAWAIHAPAI	21		6.23							3.13			
MAKUA	30		6.48							3.38			
MAKAHA	36		7.01							3.51			
WAIANAE	38		7.08							3.59			8.59
NANAKULI	44		7.22							4.14			
GILBERT	48		7.33							4.25			
EWA MILL	53	6.46	7.45		10.07	12.45			3.28	4.37		6.35	
HONOULIULI	55	6.52	7.50		10.13	12.51			3.35	4.43		6.41	
HOAEAE	56	6.56	7.53		10.16	12.54			3.38	4.47		6.45	
WAHIAWA	—			7.40			12.20	2.25			5.40		
LEILEHUA	—			8.00			12.40	2.45			6.00		
WAIPAHU	57	7.01	7.56	8.35	10.20	12.58	1.15	3.01	3.42	4.50	6.35	6.48	
WAIPIO	57	7.03	7.58	8.37	10.22	1.00	1.17	3.23	3.44	4.52		6.50	
PEARL CITY	59	7.10	8.03	8.42	10.27	1.06	1.22	3.28	3.52	4.57		6.56	9.44
WAIAU	60	7.13	8.06	8.45	10.30	1.09	1.25	3.31	3.55	5.00		6.59	
KALAUAO	61	7.17	8.09	8.48	10.33	1.12	1.28	3.34	3.58	5.03		7.02	
AIEA	62	7.21	8.13	8.52	10.37	1.16	1.32	3.38	4.02	5.07		7.06	
PUULOA	65	7.29	8.20	8.59	10.46	1.24	1.39	3.45	4.10	5.15		7.14	
HONOLULU	71	7.45	8.36	9.15	11.02	1.40	1.55	4.01	4.26	5.31	7.10	7.30	10.10
		A.M.	A.M.	A.M.	A.M.	P.M.	P.M.	P.M.	P.M.	P.M.	P.M.	P.M.	P.M.

* On Train 24, "The Haleiwa Limited," only first class tickets will be honored.
Train 32 will make no stops between Waipahu and Honolulu, except to let off passengers from Wahiawa Branch.

The Oahu Railway Timetable of 1913. *Dillingham Corporation.*

OR&L train leaving Waipahu station about 1922. *H. F. Hill photo, R. B. Jackson Collection.*

Oahu Railway Co. #32 appears on the builder's premises, the Alco-Rogers Works, in 1913. To the Salvador Railway in 1950 after many years service for the OR&L. *Alco Historic Photos — Mowhawk & Hudson Chapter NRHS.*

OR&L #32 switches box cars in the yards at Honolulu in 1946. Note braced side box cars. *Cochrane Collection, Bishop Museum.*

OR&L Co. #32 in the livery of the Salvador Railway at San Salvador station, Central America. One of four 2-8-0 machines sent to Salvador in 1950. *G. M. Best Collection.*

Above: An Alco-Manchester product (#51164—2-8-0, OR&L #34), of 1912 prepares to doublehead with OR&L #31 at the Dole Siding in 1946. Sent to Salvador in 1950. *Cochrane Collection, Bishop Museum.* Below: OR&L #35 Alco-Rogers (#55827—1916), rests in the OR&L yards in 1946. Sent to Salvador in 1950. *Cochrane Collection, Bishop Museum.*

Above: OR&L #34 leaving the roundhouse in 1946 readies for her day's work and her brakeman runs to set the proper switches. *Cochrane Collection, Bishop Museum.* Below: OR&L Co. #36, another Alco-Rogers product (#53447—1913), comes to rest under the water spouts of the service facilities at Honolulu in 1946. Also sent to Salvador in 1950. *Cochrane Collection, Bishop Museum.*

Top: Ex Oahu Railway 2-8-0 #36 appears in the livery of the Salvador Railway near the Salvador terminal about 1951. *G. M. Best Collection.* Bottom: Alco-Schenectady 2-8-0 (#46170) of 1909 vintage appears as OR&L Co. #37 at the Yard Master's office in 1946. *Cochrane Collection, Bishop Museum.*

Top: A view of the Pearl City station, Oahu Railway Co. 1946. *Cochrane Collection, Bishop Museum.* Bottom: #37 stands proudly in her stall at the OR&L Roundhouse in 1947. *Cochrane Collection, Bishop Museum.*

Top: From the Pittsburgh Works of the American Locomotive Co. comes a posed photo of OR&L #39, of 1909 vintage (#46043). *Cochrane Collection, Bishop Museum.* Center: Mr. Ephraim Shay's contribution to the Oahu Railway (Lima #3097—OR&L #44) in a builder's photo of 1920. Two of these machines (Lima #3165—OR&L #43 of 1921) were ordered and used by the railroad. Deemed unsatisfactory, generally because of their slow speed, they were finally assigned to a Waipahu siding for scrapping. *Dan Ranger Jr.* Bottom: OR&L Co. #45, Ex "Lanakila" poses on the Oahu Rwy. turntable on May 2nd, 1923. *R. B. Jackson Collection.*

In service view of OR&L Co. #39 in the Honolulu Yards 1946. *Cochrane Collection, Bishop Museum.*

Oahu Railway Co. train steps out briskly from the Honolulu Depot on April 20, 1923 with #45 on the head end. *R. B. Jackson Collection.*

A Union Pacific Railroad travel brochure on the Hawaiian Islands, Circa 1899. Comprising some 86 pages, it makes mention of the Oahu Railway & Land Company when that narrow gauge railroad was 10 years old and boasted 30 miles of track.

Top: Baldwin switcher #57 of the Oahu Railway (BLW #15336—1897), "Kalia" stands in the rail yards backed by braced-side box cars and the tanks of the Gaspro Firm in 1946. *Cochrane Collection, Bishop Museum.* Center: What was to become the pride of Walter F. Dillingham, Alco-Schenectady #66729, of 2-8-2- style, poses proudly on the builder's property in March of 1925 as Oahu Railway #60. *Alco Historic Photos—Mowhawk & Hudson Chapter NRHS.* Bottom: OR&L Co. #64 on service test after overhaul in Honolulu in 1946. A sister locomotive OR&L Co. #83 (BLW #16387—1898), was also in road service in Hawaii, but no photos appear to have survived of that locomotive. *Cochrane Collection, Bishop Museum.*

Aiea Station of the Oahu Railway. Date uncertain. *Fred Denison Collection, HSPA.*

OR&L Co. #60 stands serviced and ready for action in 1946. *Cochrane Collection, Bishop Museum.*

OR&L #60 comes "a cropper" over a broken rail at Maile Point. Date uncertain. *Cochrane Collection, Bishop Museum (#166).*

Builder's photo of OR&L Co. #64, the "Waialua" (BLW #16386—a 4-6-0 of 1898). Note the tropical shade for the Fireman later discarded when oil burners came into use. *M. H. Ferrell Collection.*

Above: OR&L #70 pauses beside the Yokohama Rubbish dump about 1947. *Cochrane Collection, Bishop Museum.* Below: Baldwin Locomotive Works, Const. No. 15388, a 2-8-0 of 1897 appears in the Honolulu yards as OR&L Co. #76 and was named "Reciprocity." *Cochrane Collection, Bishop Museum.*

The Ewa Mill station of the Oahu Railway about 1906. *Fred Denison Collection, H.S.P.A.*

One of a heavy class of four 2-8-2's which graced the OR&L Co. roster of locomotives, Oahu Railway #80 picks up her freight consist for Kahuku in 1946. *Cochran Collection, Bishop Museum.*

Above: Oahu Railway #85 an Alco-Cooke (#48585 of 1910) displays her sleek lines on the roundhouse turn table in 1946. Still in existence, she now graces the display area at Travel Town, Los Angeles, California. *Cochrane Collection, Bishop Museum.* Below: Oahu Railway #90 on the roundhouse turntable in 1939 heads up the OR&L stable of locomotives with (Lt. to Rt.) Nos. 57, 87, 37, 88, 64, 10, 98, 85, 32, 22 and 21 as spectators. *Photo Hawaii, Murray Befler photo.*

A builder's photo of Alco-Cooke #48586 as OR&L Co. #87. Of 1910 vintage note the sleek lines of this machine. *G. M. Best Collection.*

OR&L #88 demonstrates the state of the builder's art at the Alco-Cooke Works in 1916, as a classic style of narrow gauge 10 wheeler of the times. *Alco Historic Photos, Mowhawk & Hudson Chapter NRHS.*

Above: #90 as a superheated, small gauge giant, awaits her engine trials at Schenectady in 1926, as the last of a group of four "Mikados" built by Alco for the Oahu Railway of far-off Hawaii. *Alco Historic Photos, Mowhawk & Hudson Chapter NRHS.* Below: A wartime purchase from the Pacific Coast Railway, OR&L #111 held her own even with the heavy Mikados. *Cochrane Collection, Bishop Museum.*

Above: OR&L Co. #98 a Baldwin (BLW #15567) of 1897 vintage sits in retirement at Honolulu in 1946. Note stack lid to assist in keeping moisture from the smoke-box. *Cochrane Collection, Bishop Museum.* Below: A brace of OR&L locomotives peek from the roundhouse at the camera of Kent W. Cochrane in 1947. *Cochrane Collection, Bishop Museum.*

Above: On the main line OR&L M-2 waits at "Yokohama" dump to bring a gang of section men back to Honolulu in 1946. *Cochrane Collection, Bishop Museum.* Below: OR&L Co. Coach #12, one of the original Carter Bros. cars of the road. Built in Newark, California and shipped to Hawaii from San Francisco in 1889 it stands here in the depot track at Honolulu. Date uncertain. *Cochrane Collection, Bishop Museum.*

Above: OR&L Coach #9 sits on the rip track at Honolulu in 1946. *Cochrane Collection, Bishop Museum.* Below: OR&L Coach #34 rebuilt as a can car and then converted back to a passenger coach to serve the road during World War II. *Cochrane Collection, Bishop Museum.*

Above: OR&L Combine #36 stands in the Honolulu yards in 1947. *Cochrane Collection, Bishop Museum.* Below: Observation Car #64 and Coach #67 move down the line for an inspection party. Date uncertain. *Cochrane Collection, Bishop Museum.*

Above: Classic view of OR&L Co. Observation Car #64. Reported to once have carried the name "Ambassador" the car has been preserved and is currently under restoration. *Cochrane Collection, Bishop Museum.* Below: Typical OR&L so-called "long" side door Box cars of 60,000 pound capacity in the Honolulu yards 1946. *Cochrane Collection, Bishop Museum.*

Double side door Box car, OR&L Co. #5 with Caboose C-1 at Honolulu 1946. *Cochrane Collection, Bishop Museum.*

OR&L flat cars run the Wahiawa Branch loaded with pineapple in 1946. *Cochrane Collection, Bishop Museum.*

OR&L Caboose #41 on Dole line service in 1947. *Cochrane Collection, Bishop Museum.*

The Rear Brakeman rides the cupola on the OR&L in 1947. *Cochrane Collection, Bishop Museum.*

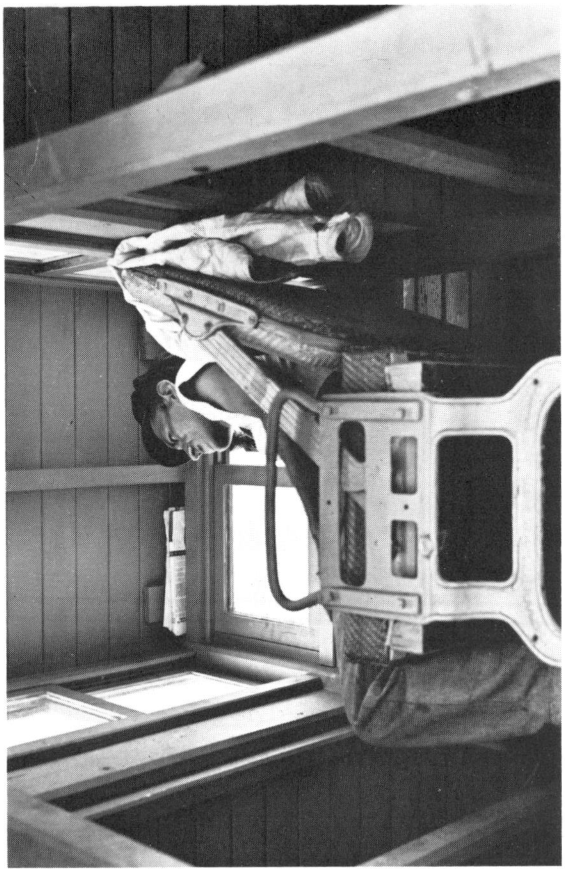

A spate of Hawaiian "liquid sunshine" prompts the Rear Brakeman of the "Extra" to Kahuku to move into the cupola of Caboose #41 in 1947. *Cochrane Collection, Bishop Museum.*

"Weed Killer" car as used by the OR&L for weed control and maintenance of way 1947. *Cochrane Collection, Bishop Museum.*

OR&L steam motive power moves "big guns" to Fort Barrette prior to WWII. Enroute one gun load shifted and all had to be reset and reblocked. Wrecking Car #45 accompanied the move. *Fred Smith, Henry Bonnell Collection.*

Above: Oahu Railway Co. #64 and train cross the trestle at Waimea Bay, Oahu, on May 18, 1923. *R. B. Jackson Collection.*
Below: M-2 of the Oahu Railway prepares to leave the Honolulu depot at Iwilei about 1930. *Author's Collection.*

Above: Dawn breaks at Honolulu as OR&L morning consists prepare to depart for Waianae and Kahuku on a winter morning. *Cochrane Collection, Bishop Museum.* Below: A Dole photo of the Oahu main line as an afternoon freight moves into the yards at Honolulu. Date uncertain. *Dole Co. photo.*

Above: Outbound to Wahiawa with empty pineapple crates, an OR&L train moves slowly from the yards at Honolulu. Date uncertain. *Dole Co. photo.* Below: G.E. Diesel, OR&L #19 assists in saving a derelict from the scrapper in January of 1965. The derelict, Hawaii Railway Co. #5, was being moved to the docks prior to her shipment to the Mainland for active restoration. *Author's Collection.*

Above: Oahu Rwy. Co. General Electric Diesel #15 stands in the yards at Honolulu as resplendent as though she just came from the builder, in 1951. *Robert Ramsay Collection.* Below: The General Electric diesel, old Oahu Railway Co. #19, is serviced atop Cumbres Pass, at an elevation of 10,022 feet in the summer of 1972. A far cry from her original tropic service, the engine is now operated by the Cumbres & Toltec railroad. *Scenic Railways Inc.*

Above: View of the OR&L rail yards at Iwilei (Lit.: The Collar-bone), taken from the pedestrian overpass between the Gaspro Plant and the California Packing cannery, at the yard entrance, Honolulu, late 1951. Punchbowl Crater can be seen in the background. *Robert Ramsay Collection*. Below: The old OR&L car repair yards with stock cars in an old shed in 1951. This area was later the site of car scrapping operations. *Robert Ramsay Collection.*

Above: A view of "Mauka" from the pedestrian bridge at the OR&L yard entrance 1951. *Robert Ramsay Collection*. Below: Veterans of the Oahu Railway stand on the rip track at Honolulu just prior to the start of scrapping in 1951. OR&L #22 was the first to feel the scrapper's torch in mid-August of 1951. *Robert Ramsay Collection.*

Little publicized work of the Oahu Railway Co. in car building at the company shops just prior to World War II. Upper left: Frame of U.S. Army Plotting car under construction in 1941. Upper right: Finished U.S. Army Plotting car GMO 1363. Started on March 31st, 1941 the unit was completed on September 30th, 1941. Weight was listed at 21,460 pounds. Lower left: U.S. Army Machine Shop Car #1A. Built by the railroad for the Ordnance Department, U.S.A., September 1941. Lower right: An oil tank car built by the railroad for the Pearl Harbor Navy Yard in August of 1941. *Ken Watson, Oahu Railway Company.*

The "Fore & Aft" of it. Front view of a brace of OR&L stable mates (Lt. to Rt.) #35, #32, #80 and #9, and rear view of the same stable mates in the service area of the Honolulu yards, 1947. *Hugh McGuire Collection.*

In better days OR&L veterans #80, Car #64, #70, #32 and #34 stand expectantly ready for service at the roundhouse in 1949. *Robert Ramsay Collection.*

OR&L Co. #43 (Lima #3165 - 1921), stands abandoned and forlorn awaiting the scrapper's torch on a Waipahu siding about 1950. *Cabral Collection.*

Above: Veteran OR&L #90 naps in the service area at Honolulu about 1946. *Hugh McGuire Collection*. Below: The Pearl City station, Oahu Railway Company, Oahu, T.H. 1946. *Hugh McGuire*.

Above: Old turntable and service section of the OR&L yards at Honolulu about 1950. *Hugh McGuire*. Below: The Pearl City station faces a now abandoned main line of the Oahu Railway in 1948. *Hugh McGuire Collection*.

Above: A locomotive frame is overhauled at the Honolulu shops of the OR&L on February 15, 1944, under the interested gaze of a G.I. railfan. *Author's Collection.* Below: Tidal wave damage on the OR&L at Mokuleia, Oahu, 1946. *James Oda Collection.*

Above: Last remnant of a mighty effort, OR&L train with observation car #64 and Coach #7 traverse the line in 1946. *Cochrane Collection, Bishop Museum.* Below: More tidal wave damage on the OR&L in 1946. This view taken at Waialua on the N.W. Oahu coast. *James Oda Collection.*

OR&L car #57 at Lualualei 1965 sent then to the Camino, Cable & Northern Railroad. This car, built on the Island in 1921, was returned to Hawaii in 1974. *Author's Collection.*

Views of the death of a giant. OR&L #60 suffers the scrapper's torch in June of 1966. *Author's Collection.*

Oahu Railway Co.'s "Leahi" passing Kahili Harbor on May 18, 1923. *R. B. Jackson Collection.*

As an "Aloha" and goodby to this sequel "Auntie" Emily Picadura bids her fondest Aloha's to a friend bound from Waianae to Honolulu in the Heyday of the Oahu Railway in 1907. *John Knaus Collection.*

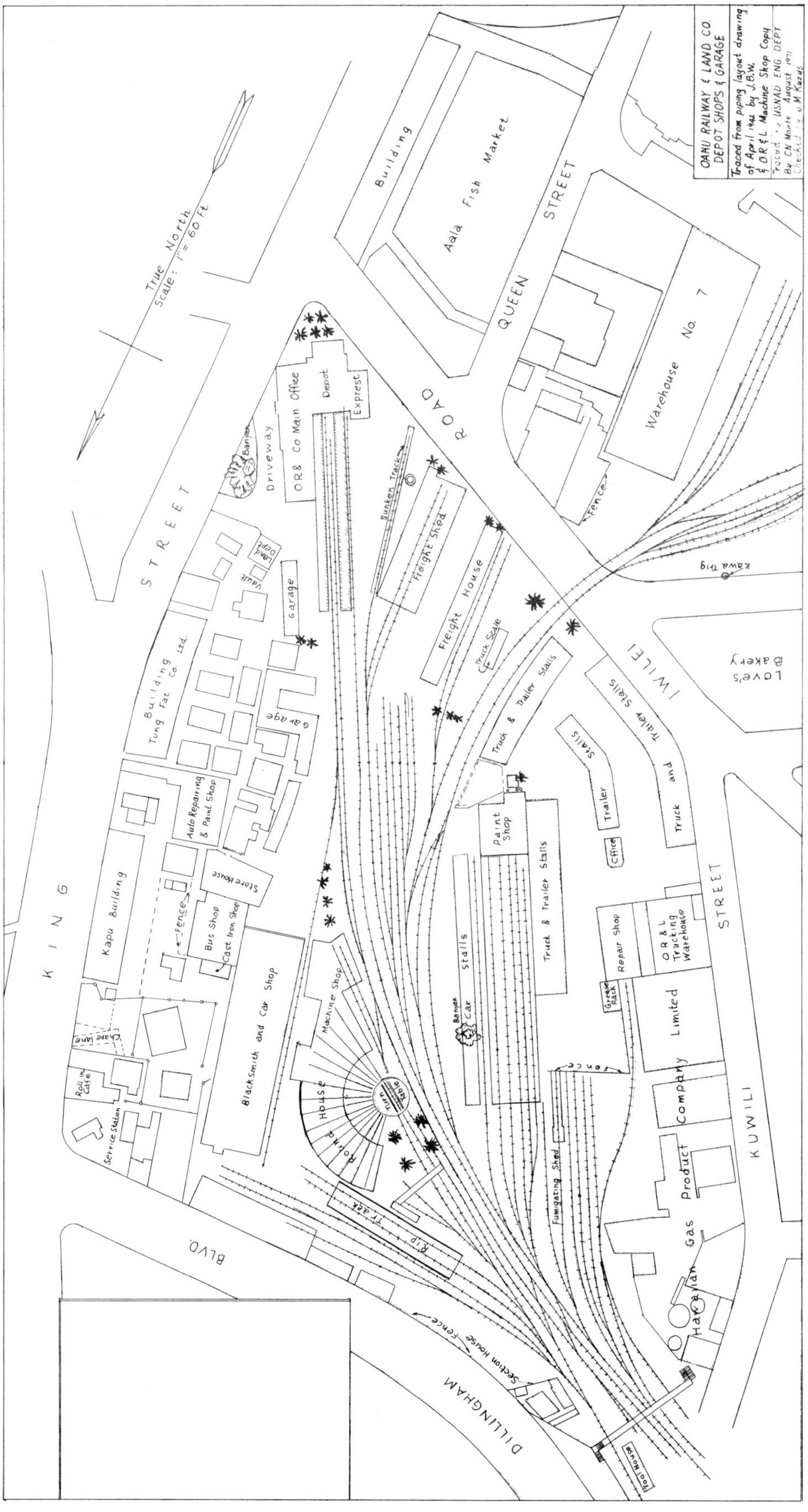

Plan view of the Oahu Railway yards, Honolulu, T.H. *Author's Collection.*

Chapter Eight /

A Latter-Day Heritage

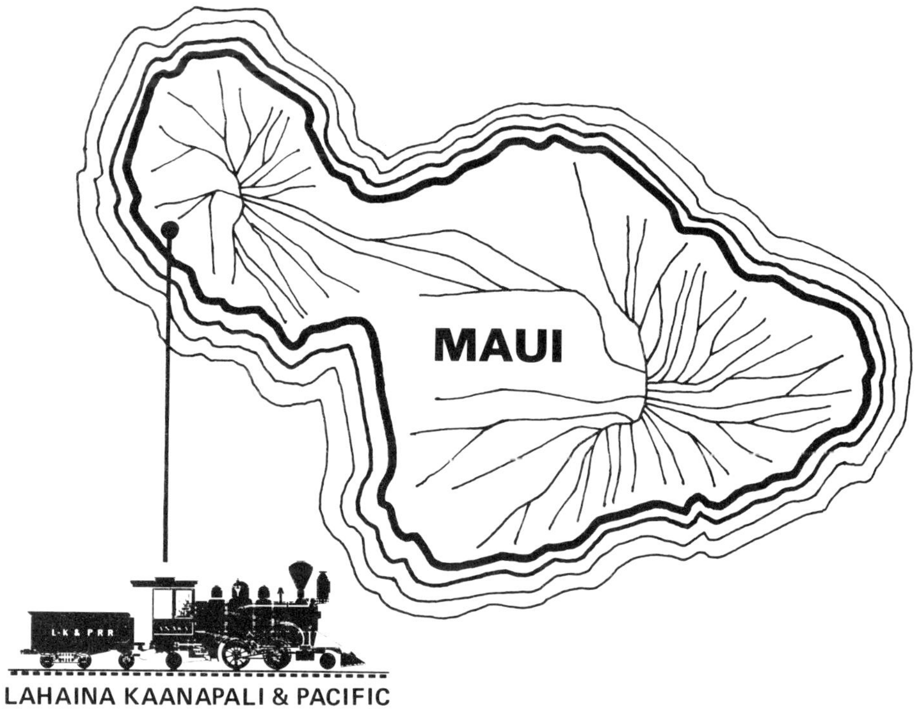

We're recreating history, with a genuine steam locomotive running between Kaanapali resort and the town of Lahaina, Maui. It's an authentic replica of Hawaii's historic 1890-1910 sugar cane trains, but this one has passenger service for the enjoyment of your clients. Stand by for ticketing . . .

ALL ABOARD!

THE SUGAR CANE TRAIN

ARRIVING SOON

Proprietors:

MAKAI CORPORATION

Sea Life Park, Waimanalo, Hawaii

Sales agents:

TRANSPORTATION CONSULTANTS INTERNATIONAL

A "Classic view" of "Anaka's" bell and sand box at Orbisonia 1970. *William Metezger.*

L.K.&P. advertising prior to the opening of the road in 1970. *L.K.&P. Railroad.*

Kaanapali Station under construction, Island of Maui 1970. *Author's Collection.*

"Anaka" as decorated with Maile Leis, traditional for new ventures in Old Hawaii. In this case the Driving of the Golden Spike, May 9, 1970. *Author's Collection.*

Views of L.K.&P. "Anaka ("Angus") #1. (H.K. Porter #7398 - 1948. Ex Carbon Limestone Co. #36 38 Inch Gauge) under test at the East Broad Top Railroad, Orbisonia, Penna., 1970. *William Metezger.*

Opposite, bottom, "Anaka" in a classic pose with side-rods down prepares to move an East Broad Top consist in March 1970.

Opposite, upper left, on her trial run "Anaka" pauses while a "hot" running crankpin is checked.

Opposite, upper right, L.K.&P. "Anaka" sits on the East Broad Top turntable in 1970.

This page, top, ready for more trial tests, "Anaka" waits expectantly on the E.B.T. turntable after minor adjustments to her running gear.

This page, below, the East Broad Top station of "Orbisonia" welcomes L.K.&P. #1 during the 1970 trials in Pennsylvania.

L.K.&P.R.R.

"Anaka" poses on the Hahakea Trestle with "Kalakaua" style car on early trials on Maui, 1970. *Author's Collection.*

Right hand close-up of "Anaka" and cars under way at speed on the Hahakea Trestle (400 feet long) in 1970. *Author's Collection.*

"Anaka" poses with L.K.&P. rolling stock, cars "Pioneer Mill," #101 and #102 "Kalakaua" while on an early trial run in 1970. *Author's Collection.*

L.K.&P. #1 rolls toward Lahaina on a typical sunny Hawaiian day with cars, "Pioneer Mill," (#101), "Kalakaua" (#102) and "Duke Kahanamoku," (#103) 1971. *Chuck Peterson, L.K.&P.*

Modern day rendition of L.K.&P. Co. Car #101 as patterned after the original old Ransomes & Rapier cars of the Hawaiian Railroad Company active on the Island of Hawaii in 1882. Named after and in honor of an early, and still existent sugar factory, near Lahaina, Maui. *Author's Collection.*

L.K.&P. #45 at the end of a day's run April 1974. *John Rutherglen Collection.*

"Anaka" and train pause at the Kaanapali Station about 1971. *Author's Collection.*

An L.K.&P. innovation, a singing Conductor. He rides the train to tell the tales of Hawaii in song and verse as well as to tout the virtues of the L.K.&P. where the name of the line is noted to be longer than its trains. Circa 1971. *Chuck Peterson, L.K.& P. Rr. Co.*

L.K.&P. Rr. #3, "Myrtle," (H.K. Porter Co. #7397 - 1947. Ex Carbon Limestone Co. #35 - 38 Inch Gauge) as rebuilt by the Keystone Light Railway Products Co., Pittsburgh, Pa. (as was "Anaka") being serviced at the Kaanapali Station, September 1973. *Robert Stout Collection.*

L.K.&P. Co. "Myrtle" with turtle back tank tender, as made from Kahului Rr. Co. #11, waits at the Lahaina Station for the run back to Kaanapali in 1973. *Robert Stout Collection.*

Side view of L.K.&P. Rr. Oahu #45, first diesel of the road April 1974. *John Rutherglen Collection.*

L.K.&P. Oahu #45 with train near Kaanapali November 1973. *Robert Stout Collection.*

View of "Anaka" and consist from under the Lahaina side of the Hahakea Trestle as "Anaka" heads for Lahaina. *Author's Collection.*

The Lahaina Station of the L.K.&P. Railroad April 1974. *John Rutherglen.*

EPILOGUE

" . . . the time may come when the riders of the High Iron on the sugar plantations of Hawaii will be legendary, and when the bells and whistles of the narrow gauge "hogs" will be stilled forever.

"And when and if that time does come, there will be recorded, somewhere, the fact that the story of High Iron in the green cane in the Hawaiian Islands provided a bright, exciting chapter in the history of railroading in America."

Carl B. Andrews, 1918

This work is now Pau Loa

My Fond Aloha to All

Jesse C. Condé